W9-AOW-104

THE RADIANT UNIVERSE

THE RADIANT UNIVERSE

ELECTRONIC IMAGES FROM SPACE

Michael Marten, John Chesterman

MACMILLAN PUBLISHING CO., INC.
New York

Produced by Quarto Publishing Ltd.,
32 Kingly Court, London, W1

Macmillan Publishing Co., Inc.
866 Third Avenue, New York, N.Y. 10022

First American Edition 1980

Library of Congress Cataloging in Publication Data

Marten, Michael.
The radiant universe.

1. Space photography. I. Chesterman, John, joint author. II. Title.
TR713.M37 1980 778.3'5 80-81397
ISBN 0-02-580420-0

The Radiant Universe was conceived, researched and written by Michael Marten and John Chesterman with the help of John Trux, Marietta Murphy-Ferris and John May of Clanose Publishers Ltd. and the Science Photo Library, London

Text: John Chesterman
Photo Research: Michael Marten; original research in the USA by John Trux; Marietta Murphy-Ferris and John May
Maps and diagrams: John Chesterman
Printed in Hong Kong

Special thanks to: Jean Lorre of the Image Processing Laboratory, Jet Propulsion Laboratory (JPL); Connie Rodriguez and Judith Anspacher at Kitt Peak; Rhea Goodwin at Hale Observatories; B.W. Hadley of the Royal Observatory, Edinburgh; Dr. Paul Gorenstein of the Center for Astrophysics, Cambridge, Massachusetts; Dr. Carl Heiles of the University of California, Berkeley; Dan Pyke and Dr. Don Wells of Kitt Peak; Charles Sheffield and Max Miller of Earth Satellite Corporation; Diane Johnson of the National Center for Atmospheric Research; Dr. Per Gloersen of the Laboratory for Atmospheric Sciences at Goddard Space Flight Center; Les Gaver of NASA's Audio-Visual Branch; Dr. Michael Abrams and Jim Soha of the Image Processing Laboratory, JPL; Jurrie J. Van der Woude, Don Lynn, Bill Becker and Dr. Kenneth L. Jones, all of JPL; H.J.P. Arnold of Space Frontiers Ltd.; and Karen Moline.

Picture Sources

Courtesy of M. J. Abrams, Jet Propulsion Laboratory, California Institute of Technology: pages 98, 99 *(bottom)*, 100, 101, 102, 103.

American Science and Engineering, Inc., Al Krieger Ph.D., co-principal investigator under the sponsorship of NASA: page 34.

© Association of Universities for Research in Astronomy, Inc., The Cerro Tololo Inter-American Observatory: pages 30–31.

© Association of Universities for Research in Astronomy, Inc., The Kitt Peak National Observatory: pages 12, 18, 19, 28, 32 *(top)*, 33, 54, *cover*.

Dr. Eric E. Becklin, University of Hawaii, and Dr. Gerry Neugebauer, California Institute of Technology: page 26.

Color photograph of Viljui River, Soviet Union, taken from the spaceship Sojus 22 with the Multispectral Camera MFK-6, manufactured by the famous optical factory of Carl Zeiss Jena: pages 86–87.

Courtesy of Earth Satellite Corporation: pages 79 *(bottom)*, 89, 95, 97, 109, 115.

Meteosat photo from E.S.O.C., Darmstadt, West Germany: pages 76, 81, 124 *(bottom)*.

Photos courtesy of General Electric: pages 79 *(top)*, 112, 113.

Dr. Per Gloersen, Goddard Space Flight Center. First published in *Polar Record*, Vol. 18, No. 116, 1977: pages 431–450, "Passive microwave images of the polar regions and research applications" by Drs H. Jay Zwally and Per Gloersen: pages 82, 83.

Dr. Paul Gorenstein, Center for Astrophysics, Harvard University: page 22.

Hale Observatories: pages 36, 37.

Dr. Carl Heiles, University of California, Berkeley: page 27.

Crown copyright, reproduced by permission of the Controller of Her Majesty's Stationery Office. Image processing by Department of Environment: pages 118–119.

Processed by the IBM Corporation under a NASA Landsat investigation. Principal investigator – Ralph Bernstein: pages 110–111.

Skylab photos. Courtesy of Itek Corporation, Lexington, Massachusetts, U.S.A.: pages 88, 108, 116.

Lick Observatory photographs: pages 8 *(top left)*, 53 *(bottom left & center)*.

Jean Lorre, Image Processing Laboratory, Jet Propulsion Laboratory; original pictures from Hale Observatories: pages 10–11, 13, 14, 15, 16, 17, 20, 21, 23, 52–53.

Dr. Freeman D. Miller, University of Michigan: page 8 *(top right)*.

NASA: pages 6–7, 9 *(top left)*, 32 *(bottom)*, 38, 41, 42, 43, 44–45, 46–47, 48, 49, 50, 51, 55, 56, 57, 58, 59, 60, 61, 62, 63, 64, 65, 66, 67, 68–69, 70–71, 72, 73, 74–75, 84, 85, 90, 91, 92–93, 94, 96, 99 *(top)*, 104, 105, 106, 107, 108, 114, 117, 120, 121, 122–123, 125 *(top)*, 126–127.

NASA/General Electric Company/Space Frontiers Ltd.: page 9 *(top right)*.

National Center for Atmospheric Research: pages 35, 39 *(top)*, 40.

National Oceanic and Atmospheric Administration/ National Environmental Satellite Service: page 80.

Photolabs, Royal Observatory, Edinburgh. Original negative by U.K. Schmidt Telescope Unit: pages 29, 53 *(bottom right)*.

United States Air Force: pages 39 *(center & bottom)*, 124 *(top)*, 125 *(bottom)*.

The photos in this book were obtained by
SCIENCE PHOTO LIBRARY
2 Blenheim Crescent, London W11 1NN.

CONTENTS

THE PLANETS 47

The third section is a roll call of high technology – Pioneer, Viking, Mariner, and Voyager – the robot observatories that have gone out into space and revealed the solar system as a more vivid, violent and complex place than anyone imagined. Not only have they rewritten the text books, but the information they have sent back is so detailed that a whole new range of sciences have been developed to interpret it.

Beyond the Cradle of the Mind 47 – The Other Planets 51 – The Comets 52 – The Thousand-Year-Old Storms 56 – High Wind on Venus 59 – The Crater Mystery 60 – The Ice Planets 63 – The Volcanoes 64 – Alien Landscapes 67 – The Landscape of Mars 68 – Sunset on Mars 70 – Earthrise from the Moon 75

THE EARTH 77

Landsat and the thousands or so other satellites now orbiting the planet have fulfilled the ultimate role of astronomy, to turn the telescopes on ourselves from the outside. The images they have given us of whole-earth systems are as beautiful as they are unfamiliar.

Decoding the Pictures 77 – Ground Truth 78 – Landsat 78 – Earth-Watch 79 – The World's Weather 80 – Meteosat 81 – The Polar Ice Scans 82 – Icefields 84 – Deserts 88 – Resources 98 – The Last Frontier 111 – The Whole-Earth Catalog 112 – Mapping the Seasons 119 – Fire and Water 116 – Southeast England 118 – The Cities 121 – The Infrared Window 122 – The Continents 124

Index 128

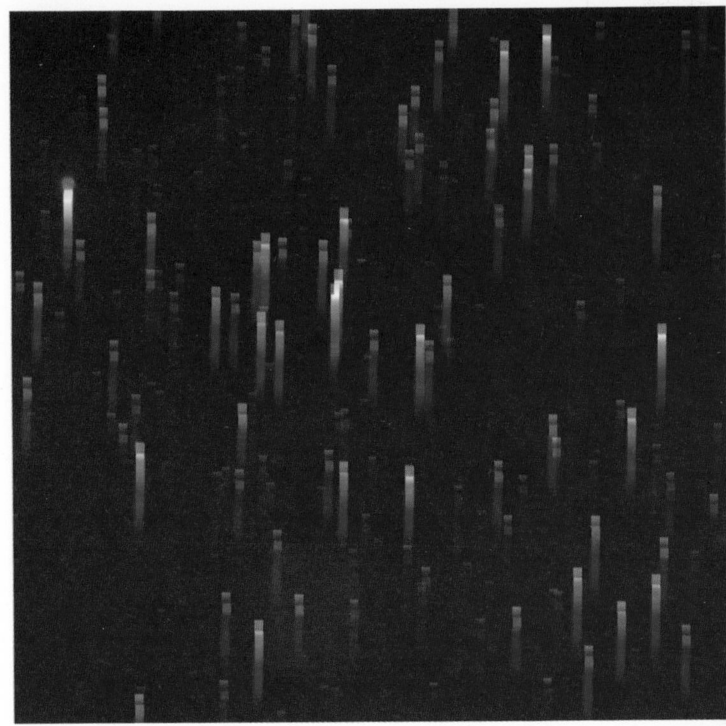

INTRODUCTION

BEYOND THE SPECTRUM

From the discovery of astronomy to the theory of relativity, our view of the universe has been marked by sudden leaps in imagination, paradigms which have suddenly widened our perspective and altered the meaning of what we thought we knew.

The trails of light apparently circling the North Star, as seen in the black-and-white photograph above, are perhaps the earliest image in human culture. The observation that in all the turning heavens there was just one star which stood still was a phenomenon which could not be explained by direct experience. It was the first step towards navigation and astronomy.

The picture was taken through a telescope at the Lick Observatory in California. It was not the stars which moved, of course, but the telescope itself, which spun on the earth's axis for a night with its shutter open. Nor does the Pole Star stand still. It just so happens that for the last few thousand years or so the earth's axis has pointed directly at it, giving us our one stable (if illusory) reference point in a shifting universe, a chance alignment which has had far reaching effects on our culture.

The picture on the right is another archetype. If the movement of stars showed us that the universe was there, the key to understanding it was the discovery of the nature of light and the seventeenth-century revolution in optics and mathematics which made it possible.

The picture was created by Professor Freeman D. Miller of the University of Michigan, who added prisms to a telescope assembly in order to turn the pinpoints of light in the Hyades Cluster into the spectrum of each individual star. It is a ground rule of astronomy that light is a code, and these distinctive signatures reveal, among other things, what each star is made of, how it burns, how heavy it is, and how fast and in what direction it is traveling.

Although light carries a significant amount of information, it is only a small part of a much larger code involving other kinds of radiation. The only reason we find light significant is that the pigment in the retina of our eyes happens to respond most effectively to radiation at wavelengths between 0.4 and 0.7 microns. In fact, light merges smoothly into the "invisible" frequencies of the spectrum – infrared, ultraviolet, X-rays and radio – all of them packed with information.

Recently, with the aid of electronic "retinae" that can "see" in other frequencies, we have started to explore the rest of the spectrum. The techniques may be in their infancy, but the discoveries they have already made are astonishing. A radio star-map, for example, shows an entirely different universe to the one we can see with our eyes. In X-ray light the surface of the sun is full of deep chasms. Patterns and structures emerge from what seemed to be empty space and mysterious energy sources are visible for the first time.

In order to translate these discoveries into an understandable and useful form, it has been necessary to invent a new language – an electronic language with electronic images.

THE ELECTRONIC IMAGE

The two images on the opposite page are pictures of the same phenomenon – the spiral cloud mass of a hurricane wheeling in toward the coast of Florida. Both were taken from space and depend on similar technology, but the chemical image (LEFT) and the electronic image (RIGHT) tell a very different story.

The electronic one is based on an infrared image from a Nimbus V satellite of Hurricane Camille, which struck

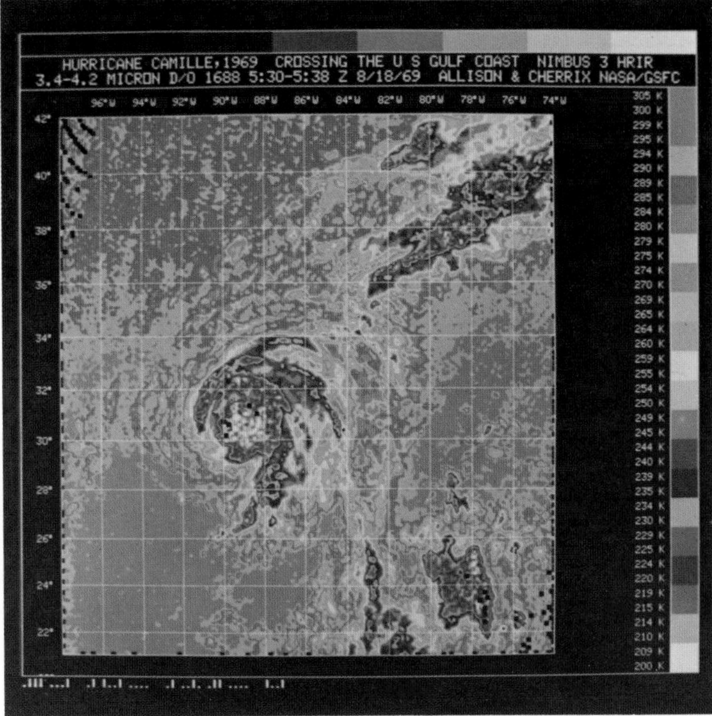

the Gulf Coast of the USA in 1969. The photograph was taken with a hand-held Hasselblad camera from the Apollo VII spacecraft, looking down from a height of 97 miles at a similar hurricane the previous year. The photograph inspires instant emotions – recognition, associative memories and adjectives like "beautiful" and "stately." It looks as precise as the other is confusing, yet the electronic image actually gives a more vivid and accurate impression of the forces at work.

Hurricanes are huge systems acting as capacitors, which pick up energy from the earth's spin and other sources, store it in steep thermoclines and pressure gradients, and then release it as the kinetic energy of 100 mph winds. The photograph shows nothing of this, and even leaves one looking for clues as to its size. The electronic image, on the other hand, shows the internal heat structure of the storm, with the levels of infrared radiation marked in colored contours, and the exact measurements of size, position and even what time it was taken included as part of the picture. It actually shows you what is happening.

Instead of using an optical lens to focus light, the electronic system uses a scanner, a device which makes a series of individual measurements of the intensity of light (or any other radiation) as it scans backward and forward across the scene. Each measurement excites an electrical impulse which can be transmitted by radio or along wires, and recorded as a string of numbers. The next stage, the equivalent to developing a photograph, is when the numerical code for the picture, including the relative position of each measurement, is fed into a computer and stored on magnetic tape. It is then possible to reproduce the picture at any time by replaying the code and turning each measurement back into a spot of appropriate brightness on a television screen.

If the image on a photographic film is faint, it is possible to increase the contrast by using special chemicals during development. In the same way, the computer can multiply the difference between measurements so that the slightest variations become apparent.

Just as color film is recorded on different layers of emulsion, so the scans of a scene made at different frequencies can be superimposed on each other to reconstruct a full-color image as realistic as any photograph. Since it is a video image, it is also possible to alter the controls, as one would those of a color television, to adjust the brightness, contrast and intensity (or saturation) of the colors at will.

And the computer can take the process of "enhancement" still further. With the original matrix of numbers safe in its memory, it can be programmed to clean up, filter, distort, exaggerate or censor the picture in a variety of ways, to extract every scrap of information from it.

There is nothing ambiguous about an electronic image, nothing hinted or implied. Nor, however far the picture is "forced," is anything invented.

It is difficult to remember this in the face of a dazzling jigsaw showing the reflective luminosity of Arctic pack ice or radiation from the event horizon of a black hole. But precision is a unique advantage of electronics and great care is taken to remove any possible artifact produced by the system itself.

Though the shapes are a form of visual algebra and their colors an artificial code, the characteristics they represent actually exist. The contours of light may be as abstract as the contours of a hillside, but the gradient is just as real.

The precise explanation for them is another matter. Not even the scientists who made them could tell you exactly what those pictures mean. We are all seeing them for the first time.

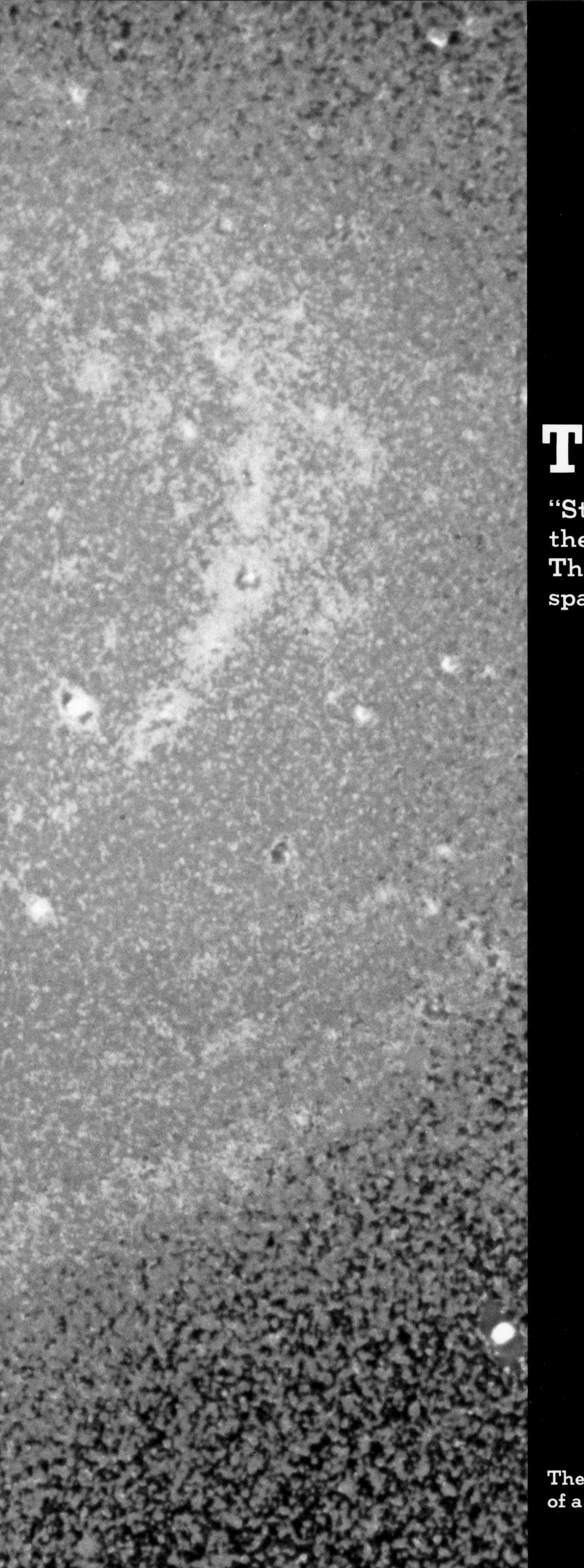

THE GALAXIES

"Stars scribble on the skies
their frosty sagas,
The gleaming cantos of unvanquished
space."

—*Hart Crane*

The golden galaxy. A computer-enhanced image of a galaxy known as M81.

The original photograph of NGC 1097, on which the electronic images on pages 13-15 were based. The details revealed by computer enhancement were all present in the photographic plates taken with the 4-meter Cerro Tololo telescope in Chile.

COMPUTER-ENHANCED GALAXIES

"You play the games," says Jean Lorre, the scientist responsible for these pictures, "and up comes the information."

It is an apt description of the mixture of luck, informed guesswork and computer programming that transformed the photograph of a galaxy called NGC 1097 (ABOVE), into this vivid explosion of color (RIGHT).

The techniques of image processing pioneered by Lorre and his colleagues at the Jet Propulsion Laboratory (JPL) in Pasadena, California, seem like a form of magic that can conjure details and color out of thin air, but, in principle, "computer enhancement" is no more than an extension of our own faculties, like the lens in a telescope.

"The information is all there," Lorre explains. "What we try to do is to rearrange it in such a way that the eye can pick it out."

The use of computers to "enhance" pictures was first developed at JPL in the late 1960s and early '70s, for the Surveyor lunar mission and the early Mariner probes to Mars, Venus and Mercury.

The Apollo spacecraft carried enough equipment to transmit live television from the moon, but the deep space probes were tiny machines, transmitting from hundreds of millions of miles away on a fraction of the power that it takes to light a Christmas tree bulb. The pictures recorded by their scanners arrived a line at a time as a string of numbers. The quality was not perfect, but since nothing could be done about the limitations of the system, ways had to be found to improve the image once it had arrived.

Using an IBM 360 series computer, now the workhorse of the space industry, JPL scientists came up with some ingenious programs to process the numerical code of the pictures. Some filtered the numbers as a linear sequence, which allowed particularly large or small objects to be selectively removed; others processed the picture as a whole, by applying algorithms and formulae such as Fourier Transformations to the numbers. Techniques of image recombination allowed individual scans to be added or subtracted from each other, and the contrast could be increased by replacing each number by its logarithmic value.

As a result of these techniques, it became possible to "clean up" pictures transmitted from space by removing random "noise" and transmission errors, enhancing the contrast and sharpening the detail, with impressive results. It was obvious from the beginning that the techniques would also have a wider application. Any image that can be scanned can be processed, and over the years the JPL computers have analyzed a range of material from medical X-rays to photographs of UFOs and the Loch Ness Monster. One of their major successes, though, has been in the field of astronomy.

Computers are a natural extension to the telescope. The high-quality photographic plates used by astronomers make ideal raw material for image processing, because they are taken at the same precisely calculated frequencies (using colored or polarized filters on the telescope) as electronic scanners. And computers, for their part, can relieve astronomers of many of their repetitive tasks, such as checking star maps.

The specialized astronomical programs devised at JPL include STARCAT, which searches the area of sky in a photograph, automatically identifying and cataloging everything it sees, and STAREM, which then goes over the picture removing all known stars and filling in the space they occupied with the average shade of the sky around them. Used together, they are a form of "instant exploration" which enables the astronomer to see whether there is anything new in the picture at the touch of a switch.

Starlight has a number of characteristics that are not immediately visible but contain hidden information about the source. Apart from the intensity of light at each wavelength, there is its saturation or the extent to which it is colored (as opposed to what color it is) and other factors, such as its hue, whether or not it is polarized and if so in what direction. The computer can map these qualities as a diagram, using contours or arrows or blocks of color; and these, in turn, can be superimposed on other images, so that a particular combination can be identified.

In order to control this bewildering range of options, a number of "master" programs were devised, including VICAR, an image-processing language that allows a picture to be held in the computer's memory while it is progressively modified by different programs.

Perhaps the most important of all the enhancement techniques is illustrated by these pictures of NGC 1097. Since the final image is displayed on a color TV screen, it is possible to replace the shades of gray in a monochrome image by a range of dazzling electronic colors.

The use of "false" color, which gives the images their surprising power and beauty, has a practical purpose. The human eye can only distinguish about a dozen levels of brightness (or intensities of light) in the average black-

The electronic tapestry. A computer-enhanced image based on an electronic scan of monochrome photographs of NGC 1097.

and-white picture, whereas it is possible for us to discriminate between literally thousands of colors (different wavelengths or frequencies). Our ability to see in color probably evolved as a protective mechanism, which enabled us to interpret the world faster and more accurately. Color makes boundaries sharper and details more precise. So by taking a single shade of gray and "stretching" it into, say, four shades of red, the range of perceptible information in the picture is extended. You can actually see more in color.

A disturbing paradox about this is that the choice of colors themselves is quite arbitrary. It is difficult to accept that the colors of a picture like the one opposite are irrelevant, but in terms of information content what matters is how different they are, not how beautiful.

Electronic imagery is one of those hybrids, born of many sciences, which have a way of challenging our basic definitions. Scientists find themselves forced to make aesthetic judgments. Astronomers, long convinced that telescopes had reached their useful limits, discover that they have been sitting on a wealth of undisclosed information, while the image processors, on whom they depend, are often at a loss to explain their own results.

Like the people who first looked at the sky through telescopes, they are simply searching for patterns, shapes, objects – anything that was previously invisible, anything new. With so many alternatives to choose from, they mix their systems like alchemists, never quite sure what they will find. Each picture is a frame of raw data to be explored in its own way.

"There's half a dozen things we know of that generally succeed," says Lorre. "Beyond that, it is a question of generating a custom process for each picture." "However," he adds, in the tone of a games player rising to the challenge, "given a bunch of pictures, I can always learn something I didn't know. Always, without exception."

The pictures on these two pages show that it is no idle boast.

THE SECRETS OF NGC 1097

The photographs show an undistinguished spiral galaxy, much like our own – one of thousands that rate no more than a catalog number. But NGC 1097 proved to be a most remarkable object. When computer-enhancement techniques were pushed to their limits step by step, extraordinary structures appeared in space around it.

The image processing of NGC 1097 was based on a series of photographic plates taken over a seven-day period with the 4-meter Cerro Tololo telescope in Chile in 1975.

The plates were electronically scanned, coded and stored in the computer at the JPL laboratories. Several plates were then combined to average out any distortion and get the best possible image.

The first picture (TOP LEFT) shows the result of the initial "cleaning-up" process, during which the center of the galaxy has been overexposed to bring out faint details in the arms. It is a marked improvement on the original, and images like this formed the basis of the color-coded version on the previous page. But it was only the beginning.

When the contrast was increased still further, a whole star field appeared in the background (CENTER LEFT). This, in turn, had to be filtered out before the process of searching the space between the stars could begin. When the computer was instructed to outline areas of equal intensity, including the tiny island of superheated gas between the spiral arms, a faint halo appeared around the galaxy (BOTTOM LEFT). Astronomers are particularly interested in studying the formation of new stars from gas and dust on the outer fringes of galaxies, and the latter picture is based on the distinctive light (in the hydrogen alpha band of the spectrum) that such areas emit. In order to make the contours stand out more clearly, the background picture of the sky was removed, revealing a ghostly image of the galaxy and its halo. The band around the galaxy does not represent a particular frequency but rather an area where there is more variety or change, in the same way that closely packed contours on a map indicate a steep escarpment.

Astronomers suspected that there were flares of radiation coming from NGC 1097, as there are from certain other galaxies, but it required electronic techniques to reveal the details of their extraordinary structure.

Further enhancement (TOP RIGHT) revealed dramatic "spikes" of radiation projecting billions of miles into space from the heart of the galaxy.

It is difficult to believe that these immensely long straight lines, one of which actually makes a right-angle bend in space, are natural phenomena, but they are simply too large to be anything else. Astronomers put them down to synchrotron radiation, a form of energy generated by electrons traveling at relativistic speeds through electromagnetic fields. They do not appear to contain any solid matter and, despite their appearance, do not go in one side of the galaxy and out the other. The jets are at different angles and are different colors.

The picture is a combination of red and blue images that have been separately processed to bring out the faintest detail in space around the galaxy. The hues have been saturated to emphasize the faint original colors, and the contrast has been increased to such an extent that bright objects, such as the center of the galaxy, are burned out. Unlike the spikes themselves, the geometric patterns around some of the stars and in the center of the galaxy, are artifacts, accidental products of the imaging system. When the process was taken to its limits (BOTTOM RIGHT), NGC 1097 revealed yet another surprise – the presence of a previously unknown fourth spike, longer and fainter than the others, to the bottom right of the picture.

Barely recognizable as a photograph any more, the contrast has been increased by so many processes that the galaxy itself has been obliterated. The blotchy background is where unwanted detail has been suppressed and replaced by the average tone of the sky around it.

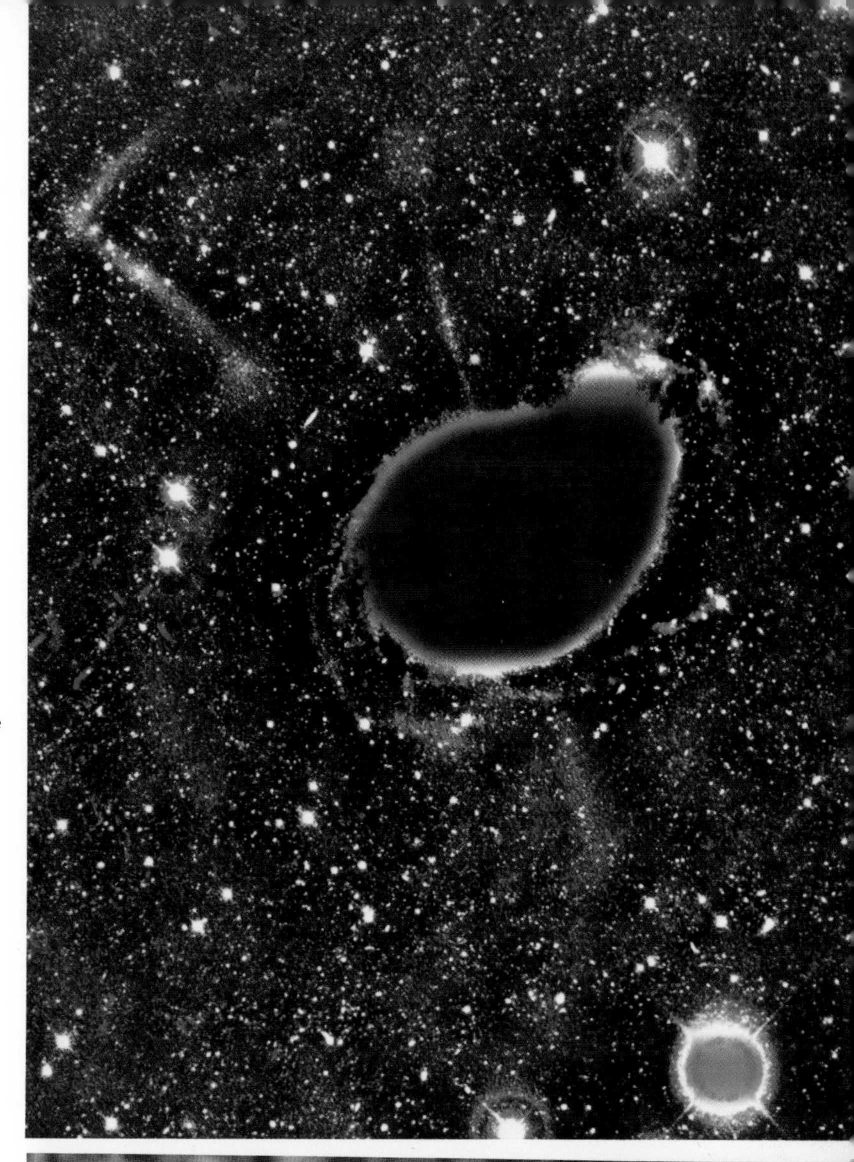

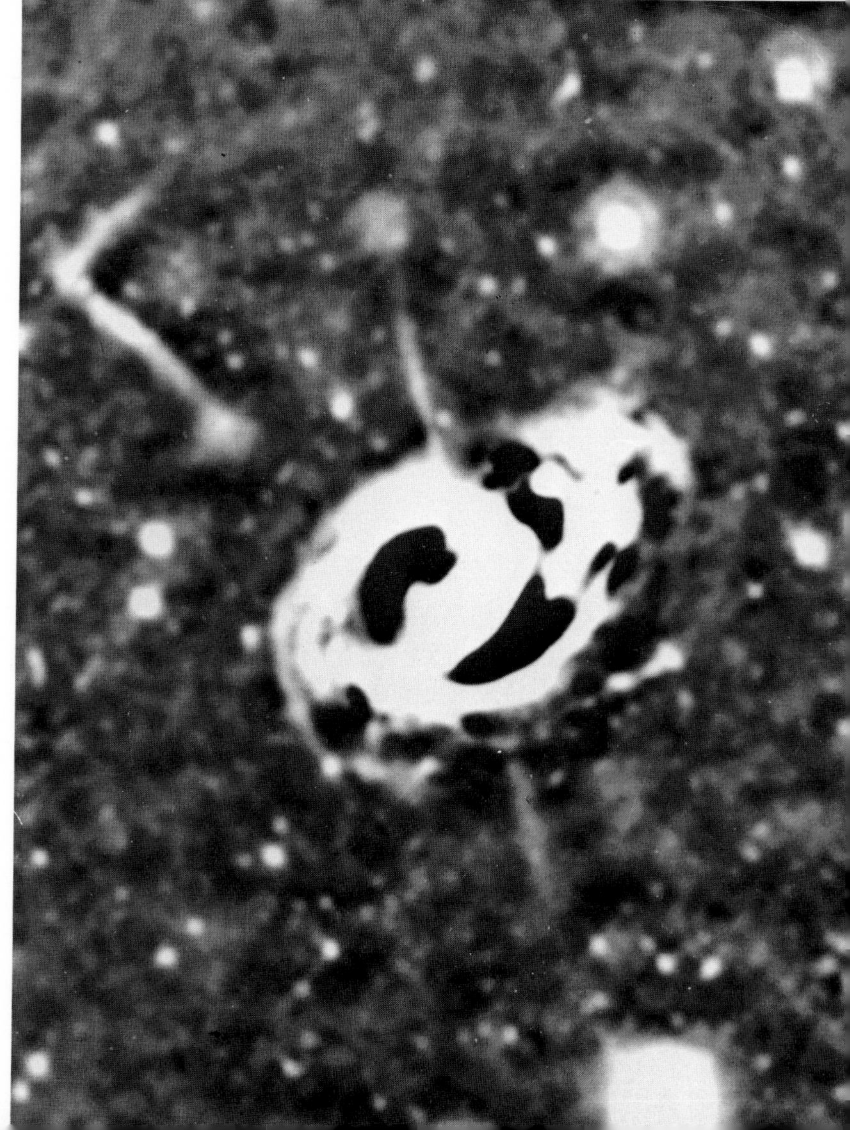

THE MYSTERY OF STEPHAN'S QUINTET

The photograph (RIGHT) shows a group of distant galaxies called Stephan's Quintet, the subject of a major astronomical mystery which computer enhancement (BELOW) may help to solve.

The relationship between the mass of a galaxy and its luminosity is well established, more or less consistent, and rarely exceeds the proportion of 1 to 10. But there are a few exceptions. In the case of the Coma cluster, for instance, it suddenly jumps to 200 to 1, and the proportions of Stephan's Quintet are even more eccentric. Either they contain an unknown source of light or an immense amount of hidden mass.

This discrepancy is overshadowed by a still more startling discovery, which seems to question the basic laws of physics. The whole edifice of modern astronomy, including the "Big Bang" theory, rests like an inverted pyramid on a formula called Hubble's Constant, which assumes that the universe is expanding and that the speed of the expansion can be measured by the distortion, or "red shift," in the light from distant objects. This enables one to calculate the age of an object (things have been slowing down since the "Big Bang") and its distance (the oldest things are most remote).

Unfortunately, objects at the same distance must obviously have the same red shift – and one of the galaxies in the picture does not. Everything indicates that it is part of the group, except that its red shift is completely different. Is the galaxy in two places at the same time? Or could Hubble be wrong?

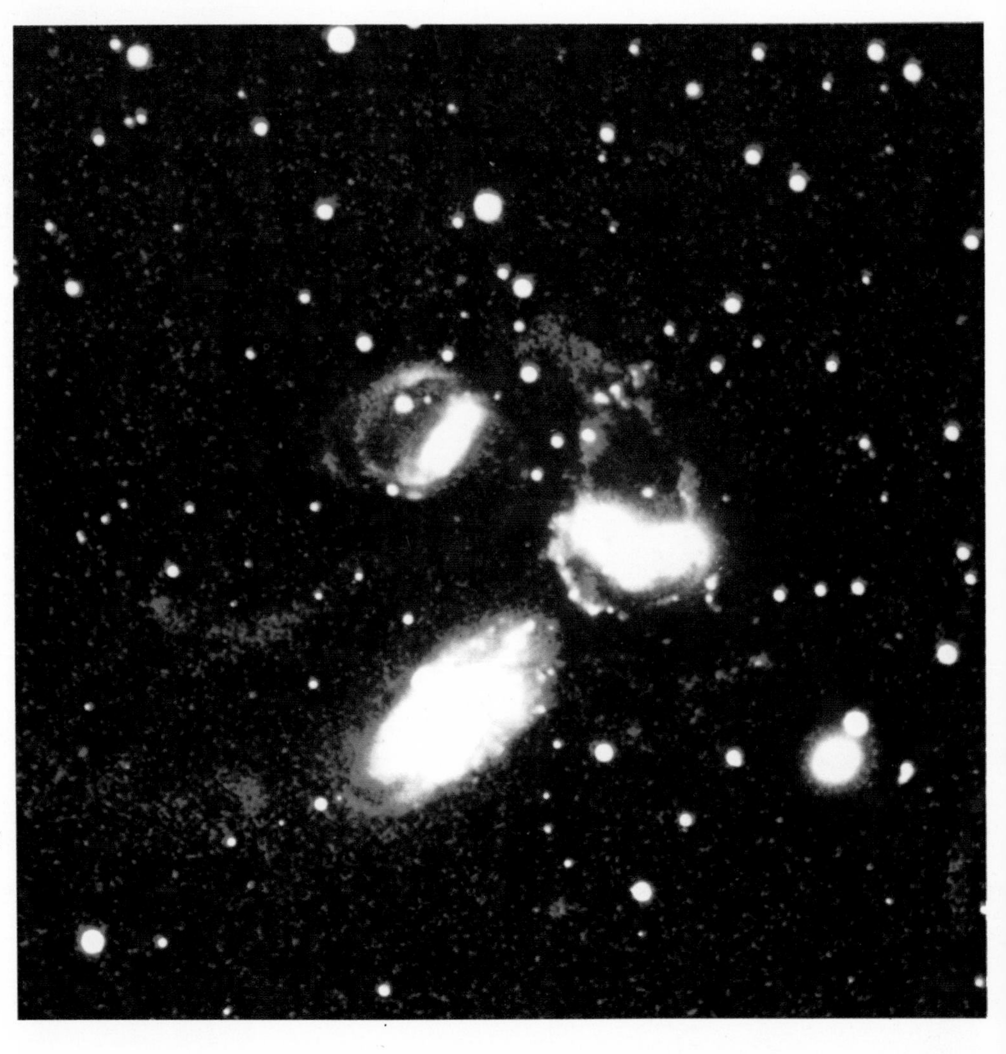

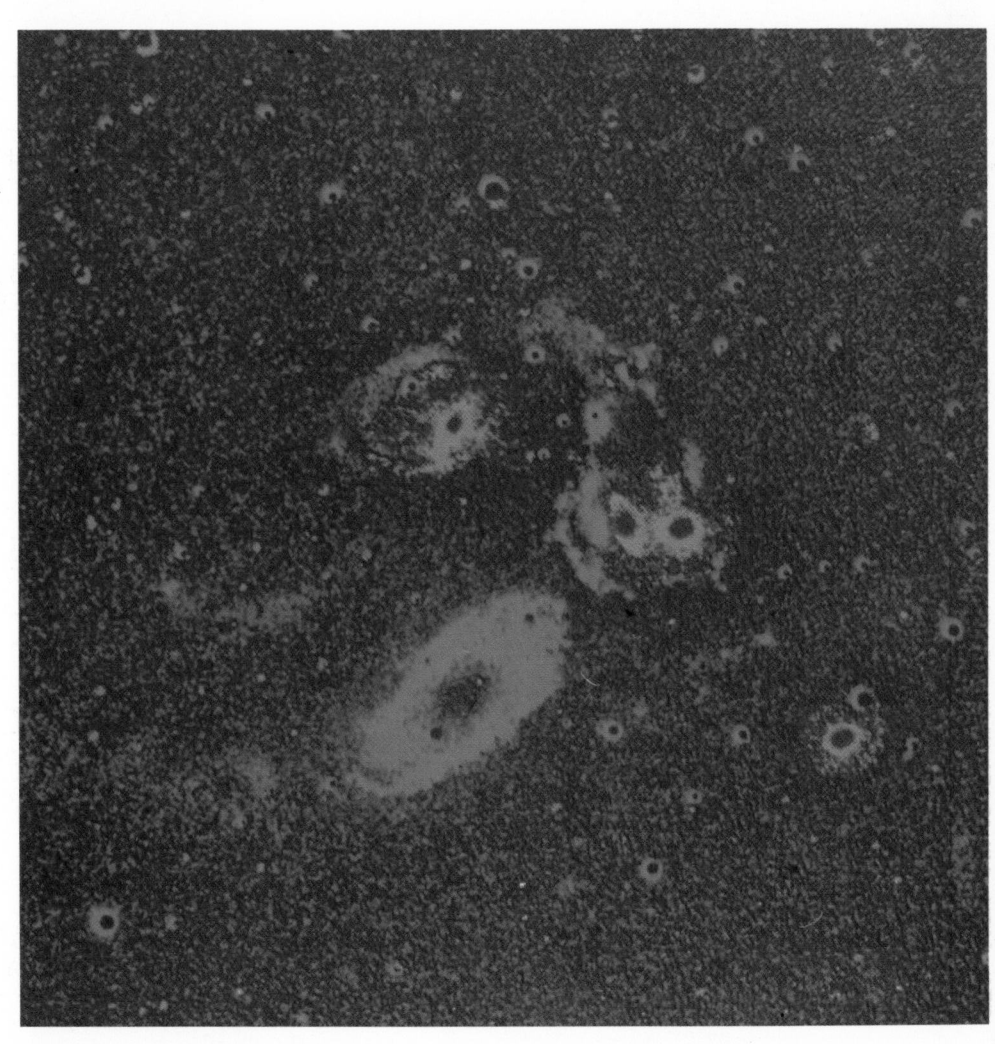

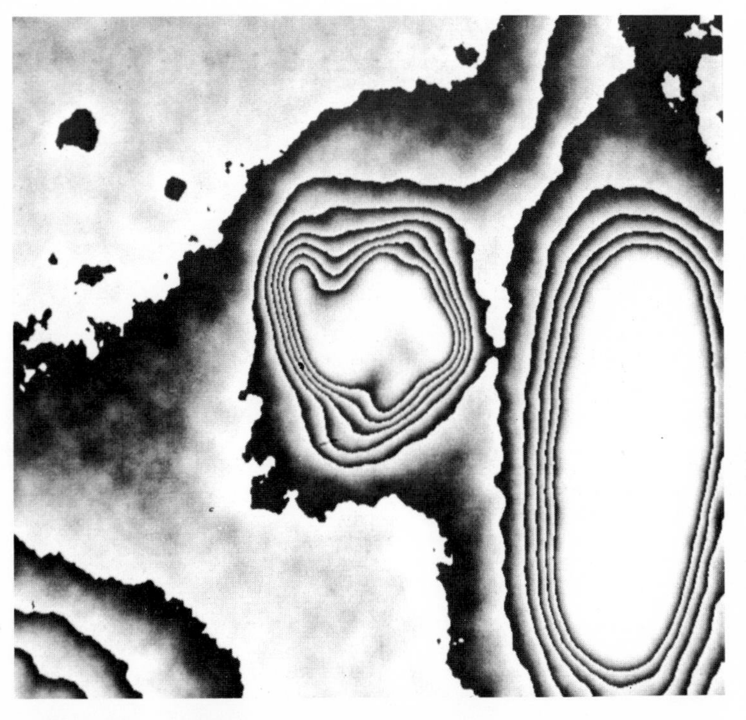

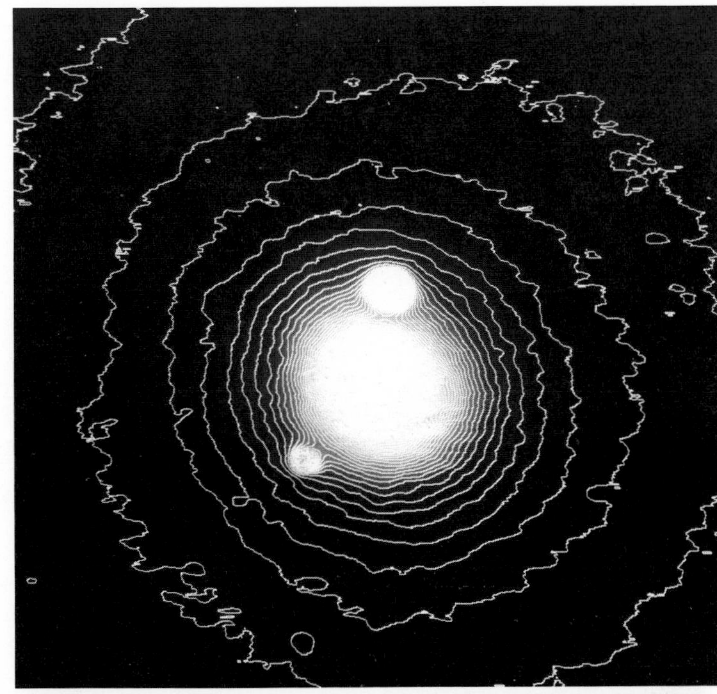

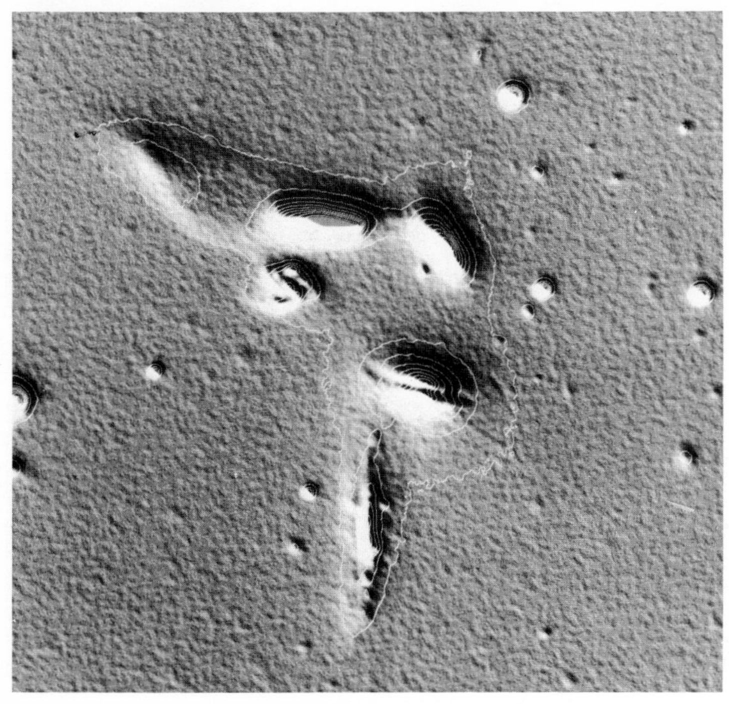

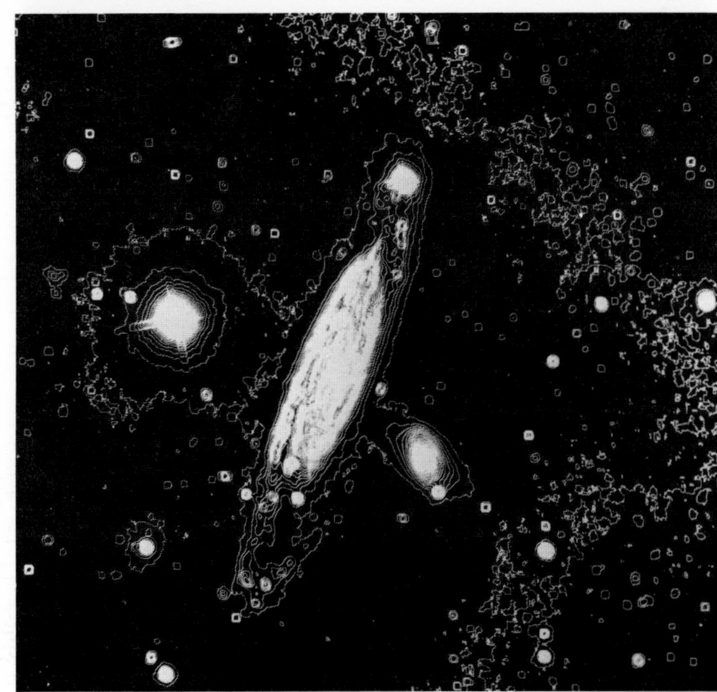

CONTOURS OF LIGHT

Radiation flows in a continuum without edges or boundaries, but the machines we use to record it work in discrete steps, making precise measurements with lines and contours. The visual shorthand that results from this compromise (such as the examples above) can look misleadingly real, and it is important to remember that they are diagrams of systems, not pictures of objects.

Paradoxically, the images are both true (they are based on actual measurements) and false (the lines do not exist as boundaries in space). Computer processing can even throw them into high relief, a step toward artificiality that makes them even more real.

The value of these diagrams lies in the way that our brains process information. The brain is divided into two halves: the left hemisphere, which analyzes linear sequences like words or numbers; and the right hemisphere, which interprets volumes and areas.

Faced with a flood of information such as the individual measurements of an electronic scan, the scientist must look for patterns to decide what is relevant and what is not. If the pixels of information are displayed simultaneously as a graph or diagram, they can be read more quickly, because it so happens that the right brain is faster at spotting patterns in pictures than the left brain is at spotting anomalies in lists.

Electronic images like these are an effective way of translating information into the right-brain language of highlights, shadows, profiles and contours.

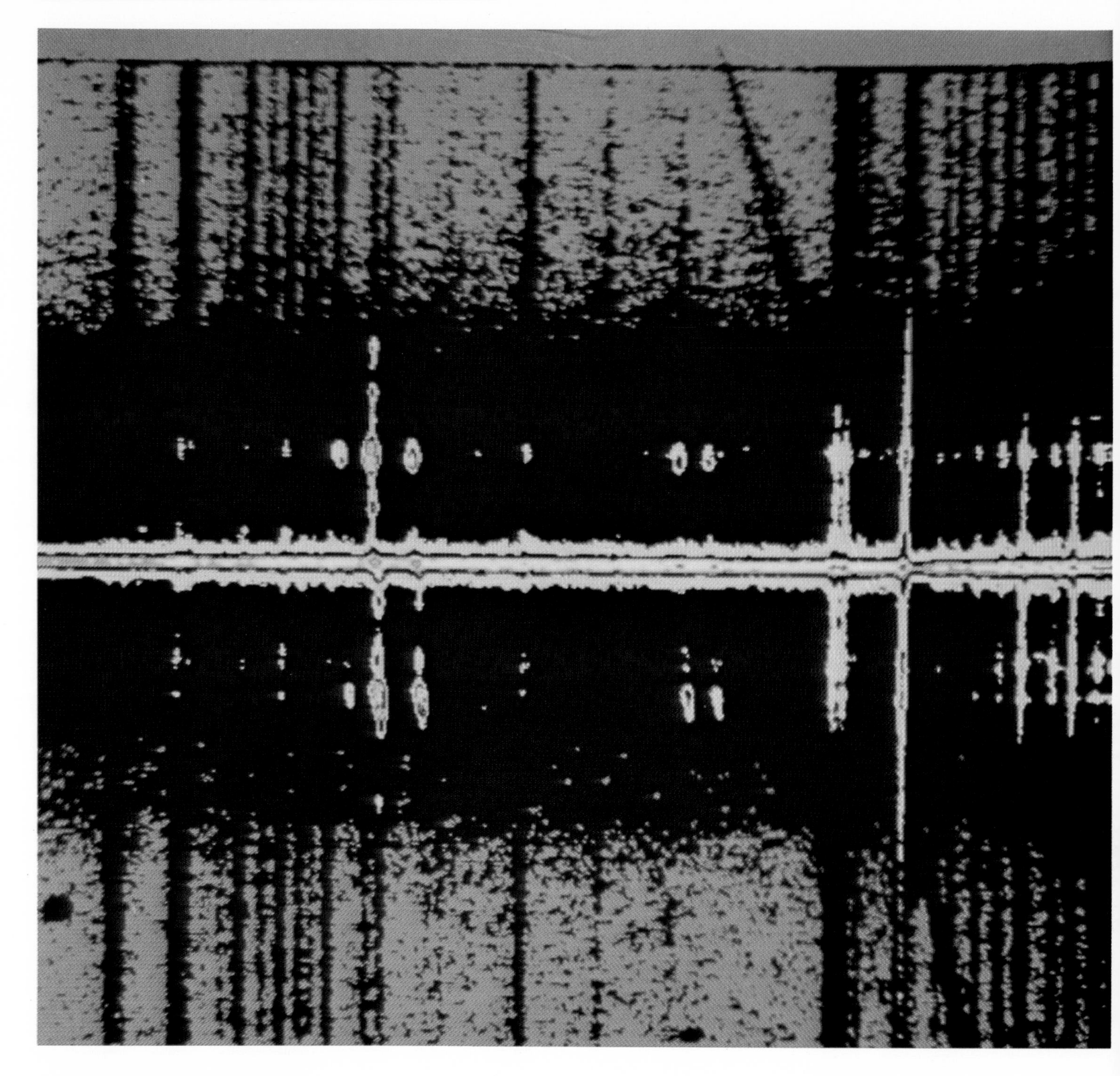

PORTRAIT OF A GALAXY

Computer enhancement can be applied to the spectrum of a galaxy, as well as to its photograph. In this case the two images make up a portrait of NGC 5364, a galaxy much like our own, about 75 million light years away.

Galaxies come in all shapes and sizes, from ragged little groups to gigantic hazy spheres containing 10 trillion star systems. But this is the most common type, a collection of about 100 billion stars in the familiar shape of a flattened disc with arms radiating from a central nucleus.

Galaxies are a fairly recent discovery. It was not until 1927 that the first objects were identified outside the Milky Way, but they have since become a major target of research.

The reason astronomers are so interested in them is that unlike stars, which are temporary phenomena, the evolution of galaxies reflects the history of the universe itself. On current estimates, the universe began with the "Big Bang" about 16.5 billion years ago, and the galaxies formed 4 or 5 billion years later, as the expanding cloud of gas started to break up.

The early galaxies were very big and bright, and with so much material available that the stars were formed at up to 3,000 times their present rate.

Just how these enormous, unstable systems evolved into the shapes we see today has still to be worked out in detail, but recent discoveries have offered clues.

Galaxies "grow" by a special kind of chain reaction, with old stars exploding as supernovae every century or so, and several new ones being formed each year. At the same time, the spin sets up "density waves," which help the formation of new stars and separates the material into arms or ridges. In this way, although the mass remains more or less the same, the material gradually becomes more concentrated and the structure more defined.

Left to themselves they would all become spirals, but there are other forces at work that dramatically change their shapes. It is now known that galaxies are constantly interacting and colliding with each other (see pages 24-

25), adding or subtracting hosts of stars each time they do so. When these crashes were simulated on computers in an accelerated form, it was found that almost any shape could be accounted for by collisions between the original spiral discs.

Other discoveries, such as the immense energy systems in the cores of galaxies, present more of a problem. As the following pages show, some of these events are on such a scale, and are so violent, that it is impossible to even guess what is happening.

In fact, the dense nucleus of a galaxy may prove to be as far beyond our experience as the nucleus of an atom – and an even greater challenge to science.

ABOVE: **This photograph of the NGC 5364 galaxy has been enhanced to show the areas of star formation. These form a red pattern around the bright nuclear bulge, with small pockets strung out like beads along the arms of the galaxy.**

OPPOSITE: **The spectrum shows the light from stars in the galaxy's nucleus as a yellow line along the middle. The bright irregular spots on either side are due to emission lines from nebulae in the spiral arms. Night-sky emissions (the faint glow of the earth's atmosphere) appear as thin vertital lines across the spectrum.**

The original spectrograph and photograph were both made with the 48-inch Mayall Telescope by Drs. Jean and Larry Goad and processed at Kitt Peak, the first major observatory with its own image-processing laboratory.

THE EXPLODING GALAXY

The explosion tearing the heart out of the M82 galaxy was the first disaster of its kind to be discovered. The photograph (ABOVE), taken with the 200-inch Mount Palomar telescope, and the computer-enhanced image by Jean Lorre (RIGHT), show an event of unimaginable violence which began 2 million years ago and is still giving off more infrared radiation than we receive from the rest of the Milky Way. The tendrils of gas are expanding at 600 miles (960km) a second and are now about 1,400 light years long.

The idea of exploding galaxies was first put forward, almost as a joke, by scientists looking for the origin of cosmic rays, the high-energy particles that come at us from every direction in space.

It began to be taken more seriously in the 1950s when radio signals were found to be coming from areas of sky where there were no stars. It seemed to be confirmed when the first visible objects to be identified with the signals turned out to be some of the faintest and most remote galaxies. Because the signals came from such a great distance, the original source, like that of cosmic rays, must have been extremely powerful.

The only process known to physics that could explain them was synchrotron radiation, an effect discovered in atom-smashing machines when electrons were accelerated to almost the speed of light in strong magnetic fields.

But the energy required to produce synchrotron radiation on this scale was immense. The nuclear fusion that takes place in stars and H-bombs was not powerful enough. In fact, calculations showed that it would require the explosion of an entire galaxy.

There was still no evidence of what was happening in the galaxies, because they were 700 million light years away, at the extreme limit of optical telescopes.

The breakthrough came in 1961, when similar signals were found to be coming from M82, a neighboring galaxy only 13 million light years away and well within the range of telescopes.

The full resources of astronomy were turned on it. The Mount Palomar telescope was used to photograph it at different frequencies, the latest chemical techniques were used to enhance the plates and they were carefully measured, compared and superimposed. The results, though not as detailed as later electronic images, were probably the last great triumph of traditional astronomy.

Whatever was happening in M82, it was the largest

explosion ever recorded and unquestionably powerful enough to produce cosmic rays. The streamers of gas were identified as ionized hydrogen, and the light from them was polarized which was another symptom of strong magnetic fields. The evidence of synchrotron radiation seemed conclusive.

Some of these conclusions have since been challenged. For instance, synchrotron radiation should only affect the light from hydrogen, and it does not explain why the whole spectrum of M82, including the light from nitrogen, sulphur and other chemicals, is also polarized.

This is only one of many anomalies and recent discoveries about M82, which have put it firmly back in the field of speculation.

Instead of a bright nucleus of stars, there appears to be nothing at the heart of the galaxy but empty space, glowing with infrared radiation. In fact, there are very few stars to be seen anywhere in the galaxy, because it seems to consist mainly of dust. This is another mystery because the clouds of dust are too large and moving too slowly to have been produced by the explosion itself – if, indeed, it is an explosion. And if the dust has always been there, why have no stars been formed out of it?

The radio signals have been found to come from two sources, one on either side of the nucleus, which is often the sign of a "black hole." Is this what is causing the disturbance? Could the dust be absorbing its radiation and re-emitting it as infrared? And if the magnetic field was strong enough, could the individual particles of dust be lined up, like iron filings, so that the whole galaxy would act as a polarizing filter for its own light?

The theories multiply, as the object that promised to be the test bed of cosmic physics turns out to be one of its most mysterious problems.

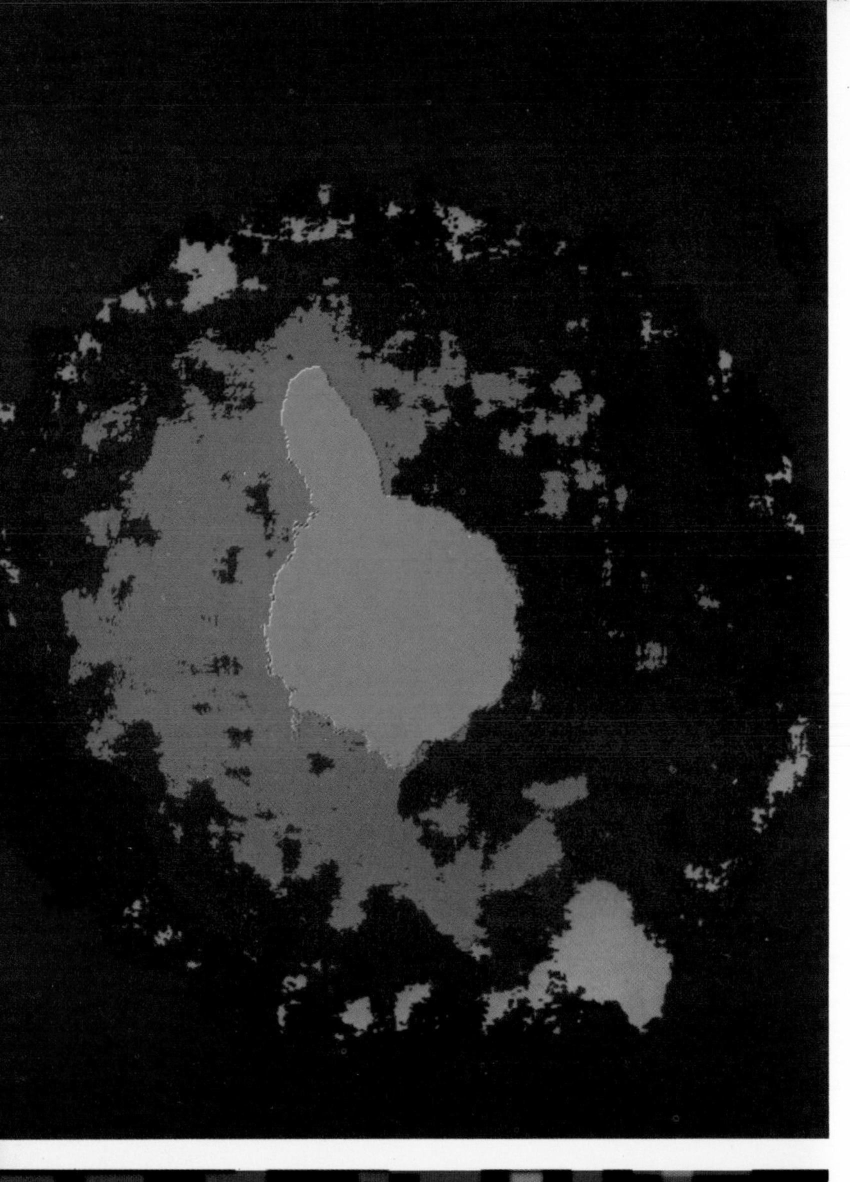

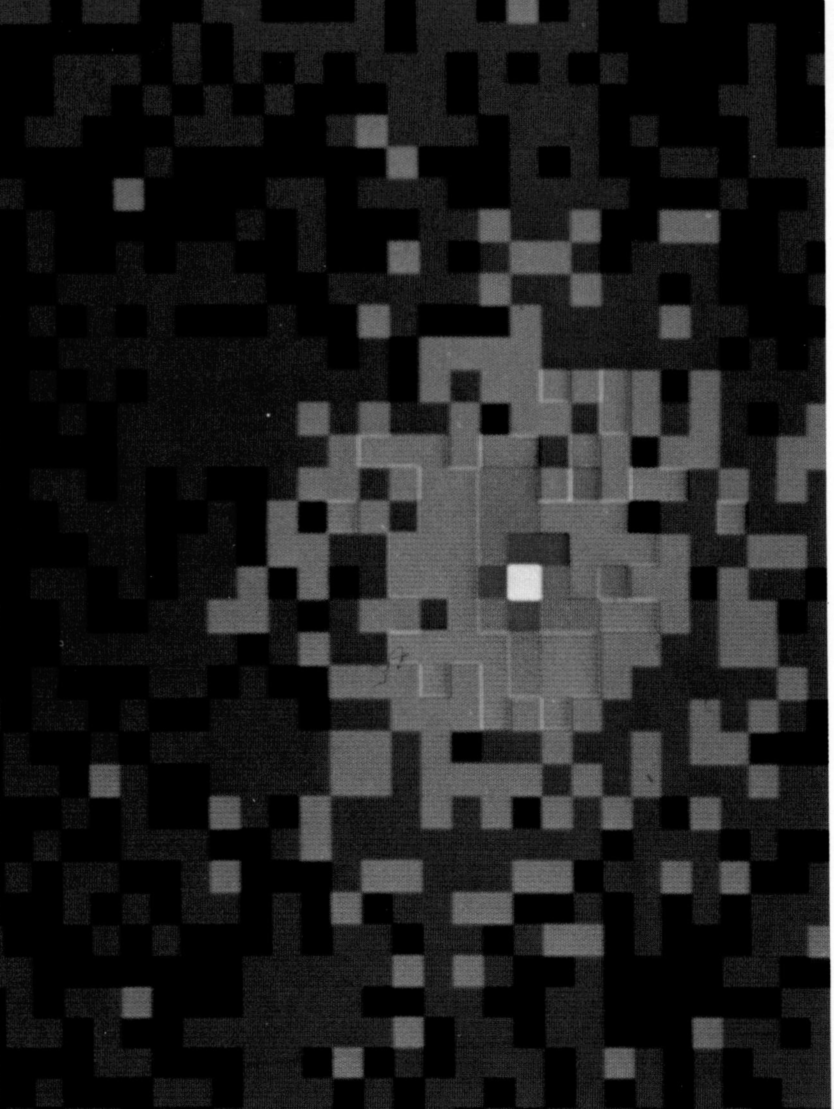

THE BLACK HOLE OF M87

This is the largest galaxy so far discovered – a huge, egg-shaped assembly of about 10 trillion stars called M87 – and the events taking place within it are on an appropriate scale.

It is relatively close to us and is situated in the constellation of Virgo, about 10 million light years away. But there the resemblance ends. There is very little dust or gas in M87, as if the galaxy had transformed all its raw material into stars and planets. This may even be the case, as the orbits of stars in this kind of elliptical galaxy take them on repeated sweeps through the center, instead of being kept at a discrete distance as they are in spiral galaxies like our own.

The most distinctive feature of M87, first discovered in 1918, is the vivid blue jet of material apparently being spewed into space. It is pointing slightly toward us, so the length is foreshortened. There may be a similar protuberance on the other side, pointing away from us.

The computer-enhanced image (OPPOSITE) shows its structure in some detail. Like the "spikes" of NGC 1097, the jet is evidently self-powered and must involve some form of natural amplification. If all the energy that gave rise to it came from the nucleus, we would not be able to see it. This is because the jet is over 6,500 light years long, and a particle that had traveled that far would no longer have enough energy to emit light.

Once again synchrotron radiation appears to be the only explanation. But unlike M82, something has been discovered in the center of the galaxy which could be causing it – the "black hole," revealed by X-ray maps (LEFT).

The characteristics of black holes have been worked out in theory, and several possible candidates have been identified, but this is the nearest to a picture of one which has ever been obtained.

It is impossible to observe these elusive objects directly. Their gravity is so powerful that not even light can escape from them, so they are literally invisible. The only physical clue to their existence are the intense X-rays emitted as material is sucked in from the space around them.

This image is an X-ray picture taken above the earth's atmosphere by a camera mounted on a sounding rocket and subsequently processed by computer. It shows an area of sky representing an arc of about one degree, in the region of Virgo. There are many galaxies, and countless billions of stars, in the region, but they are overwhelmed by the glare of X-rays from a single brilliant source. The focus of this – the white square in the center of the picture – is M87.

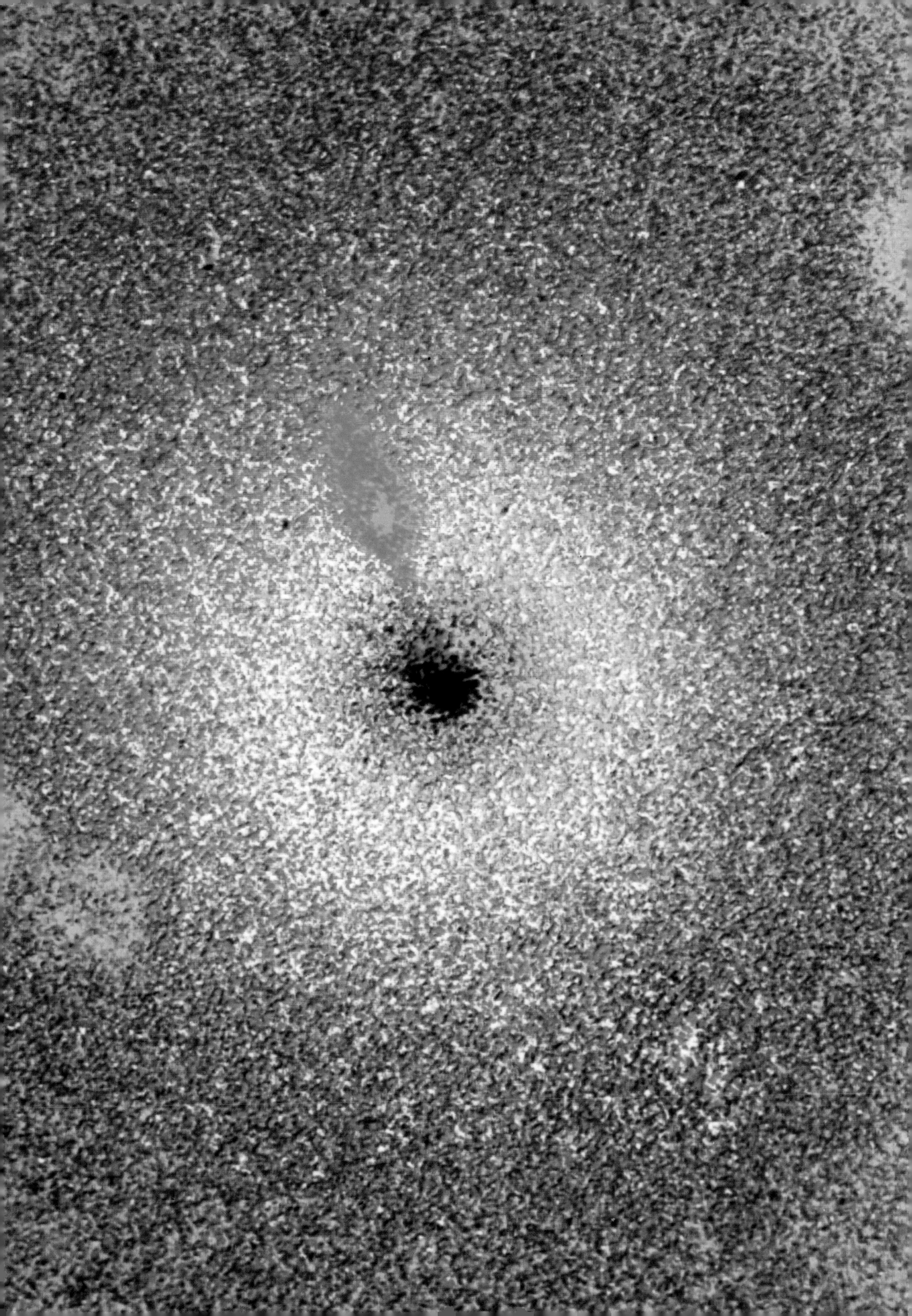

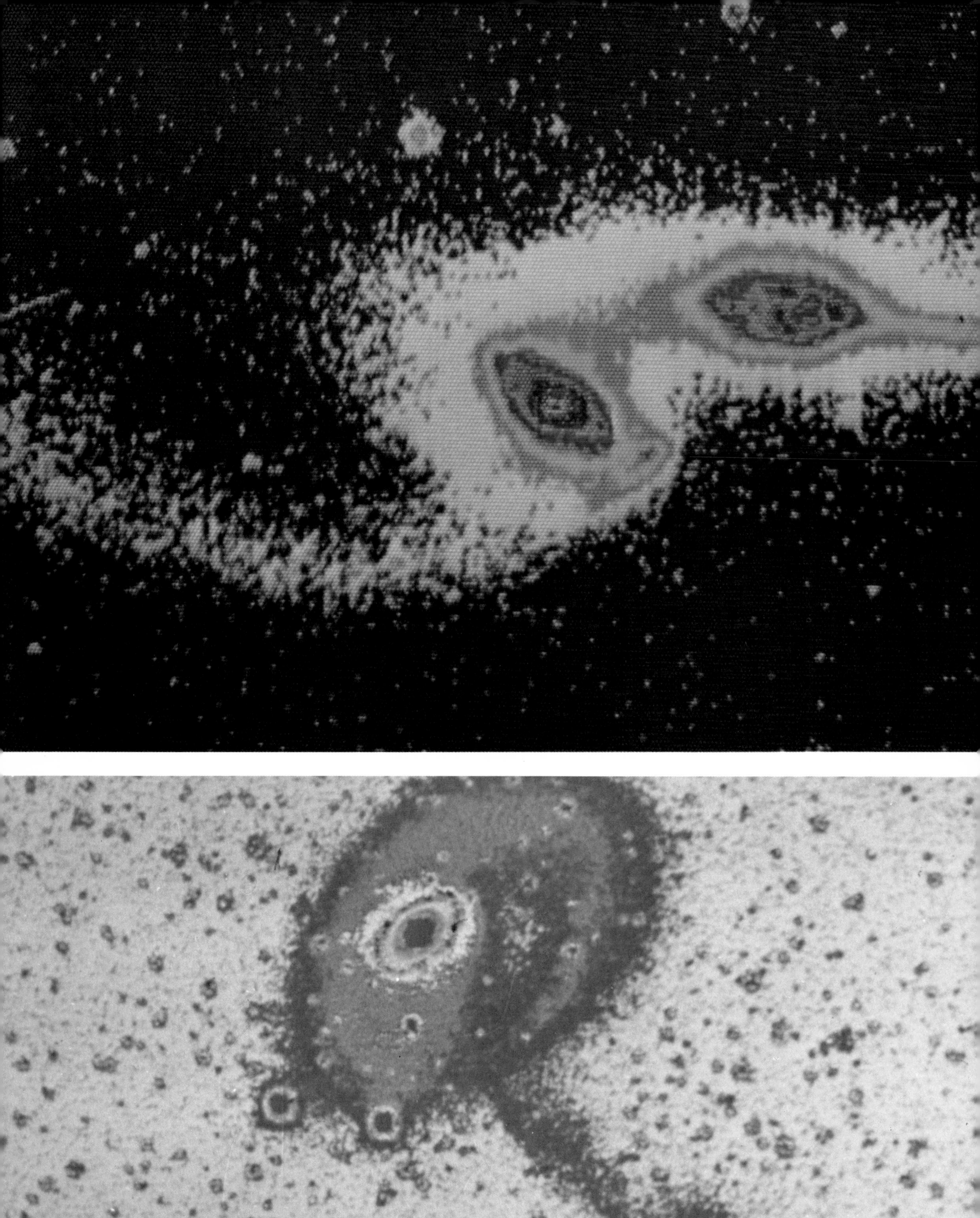
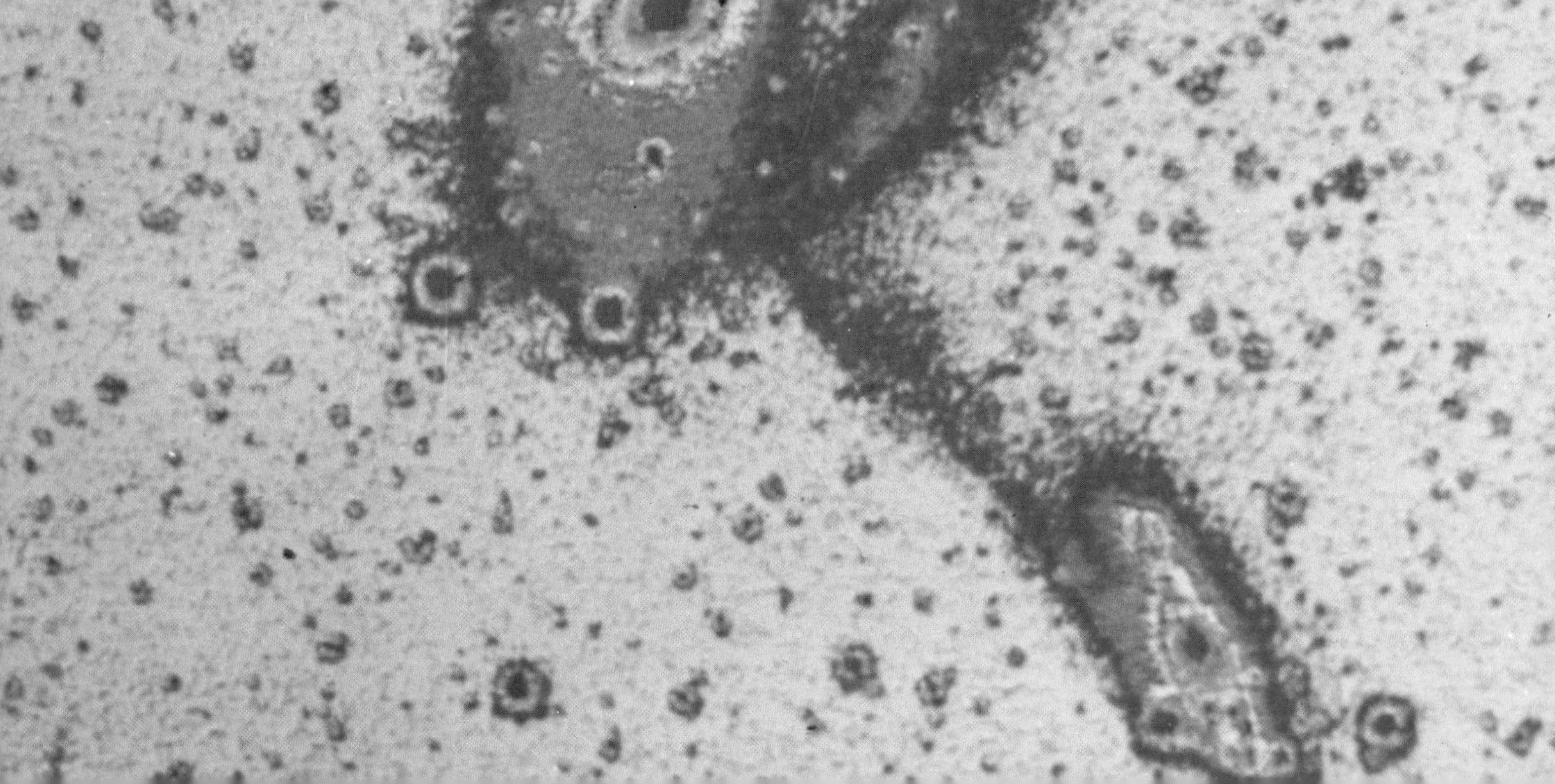

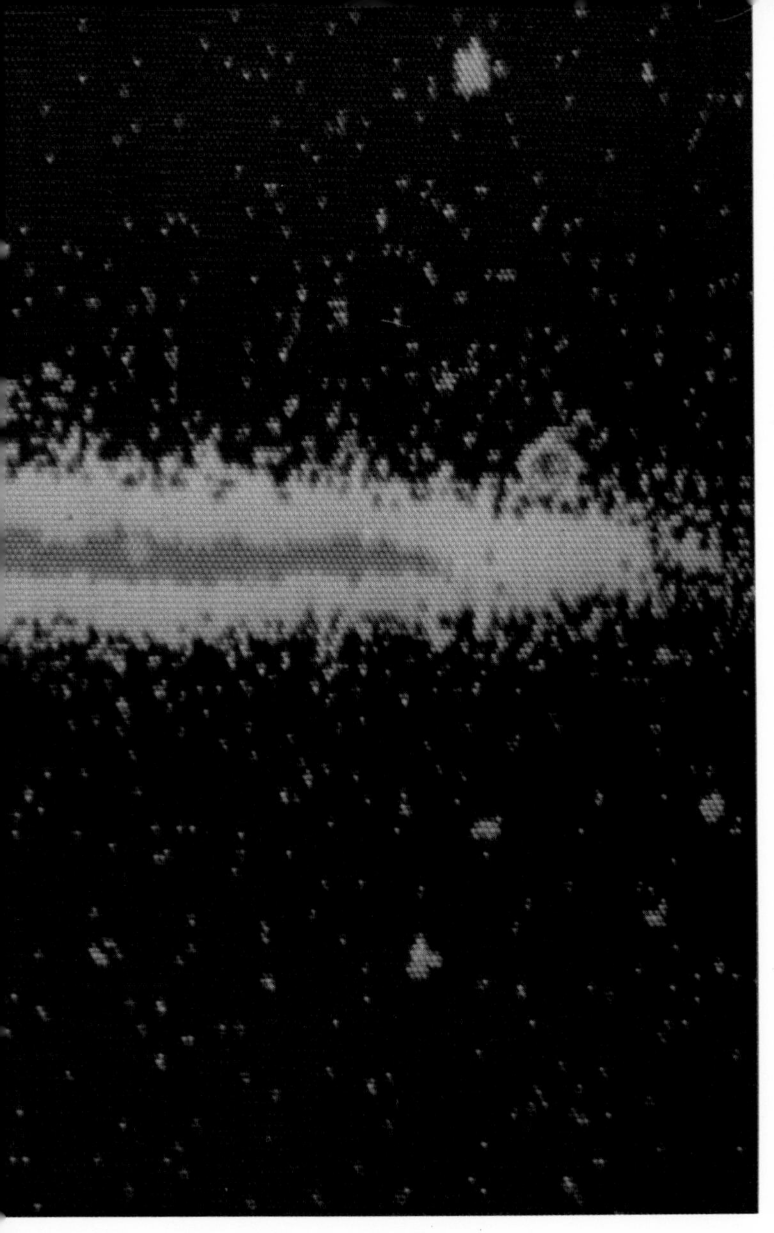

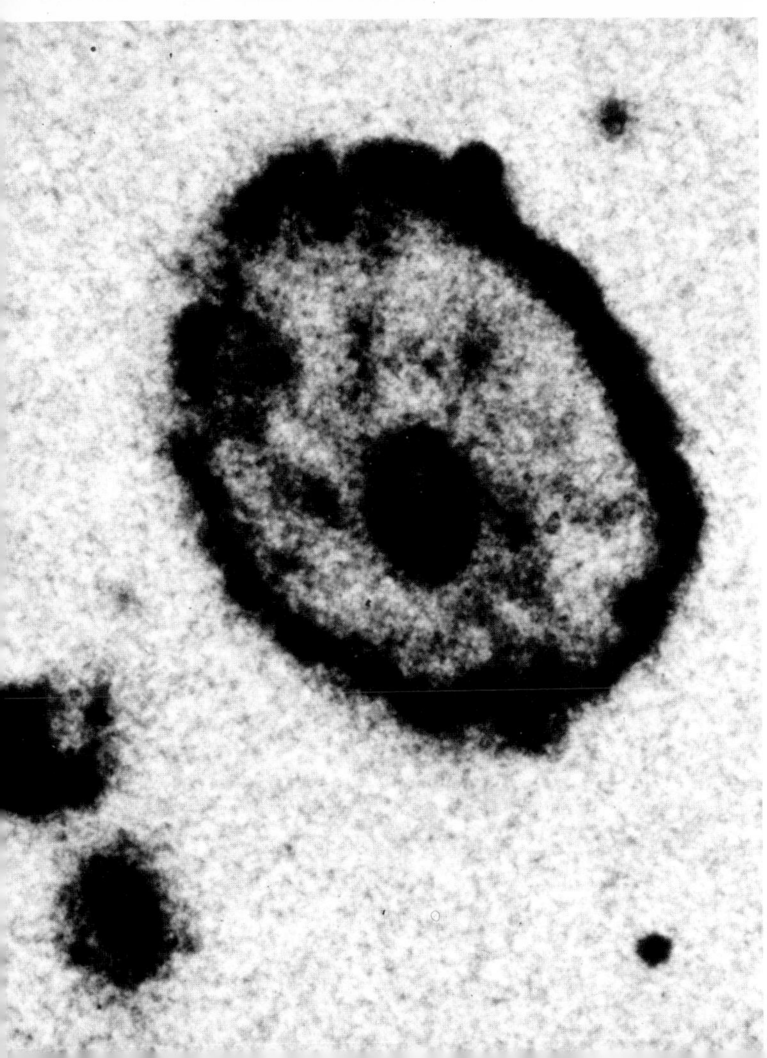

COLLIDING GALAXIES

In a constantly shifting universe one would expect there to be accidental collisions between stars. But they are an immense distance apart – about 30 million times their own diameter – and collisions between them are so rare that not one has ever been recorded. Galaxies, on the other hand, are relatively close together – about 150 times their own diameter – and collisions between them are common.

The chances of a collision have been estimated at about 25 to 1, and where galaxies are clustered together each one stands a better than even chance of undergoing at least one collision during its lifetime.

If galaxies are traveling side by side at the same speed, the encounter can be relatively tranquil. The Milky Way, for instance, is colliding in this way with one of its satellite galaxies, the Magellanic Cloud, right now.

Since the stars themselves are unlikely to collide, the galaxies should, in theory, pass right through each other, even when they meet head on. But there is evidence to the contrary. The stars might miss each other, but the nebulae within each galaxy are bound to collide, producing clouds of dense, unstable gas and shock waves. The conditions within each galaxy would become so unstable that the orbits of stars would be warped and planetary orbits would disintegrate, while compressive forces would raise the temperature and radiation levels. The process is a slow one, but life forms on planets anywhere near the point of contact would be unlikely to survive. Even when the galaxies pass each other at a distance, the massive tidal pull of their gravity distorts their shapes and siphons off streams of material from one to the other.

The photograph (BOTTOM, FAR LEFT), taken at the Cerro Tololo Observatory in Chile and computer-enhanced at Kitt Peak, shows a bridge of this sort developing between star hosts more than five million light years apart.

The galaxies in the picture (TOP), taken through Kitt Peak's 4-meter telescope and color-coded to show levels of brightness, have been overwhelmed by an even more extreme catastrophe. The tails of material streaming away from both galaxies are so elongated that scientists at Kitt Peak have called them "The Mice."

The photograph of an exploding "ring" galaxy (BOTTOM, LEFT) shows one of the worst disasters of all – a head-on collision between dense star clusters in the center of two galaxies. One has burst outward in an expanding circle, as the other passed right through it and carried away the entire nucleus. Giant elliptical galaxies, like M87, are thought to be built up in this way by a series of collisions.

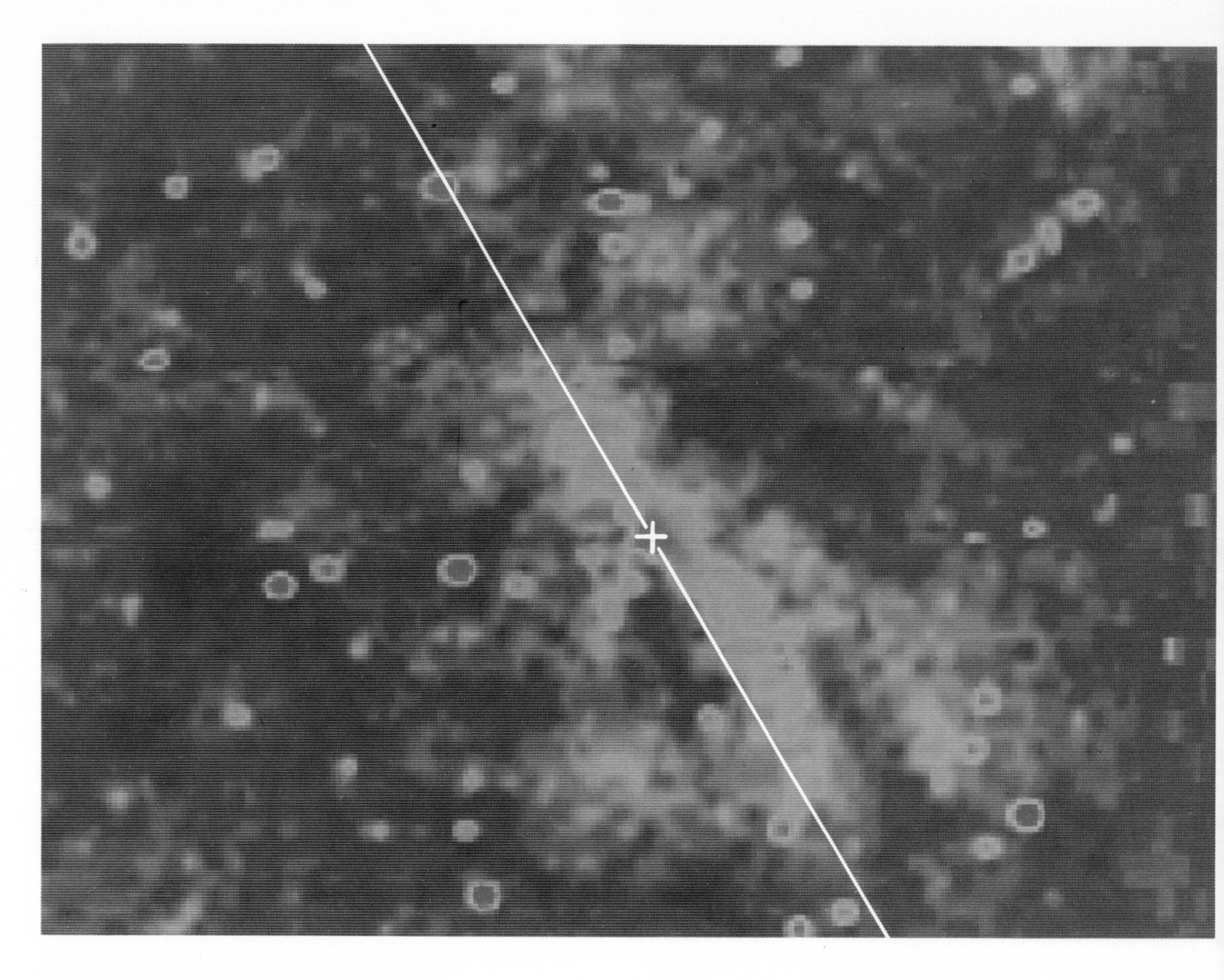

THE CENTER OF THE MILKY WAY

This is the first picture ever made of the center of our own galaxy – part of new research that indicates that, like M87, there may be a massive black hole in the heart of the Milky Way.

It is a view once thought impossible, looking 30,000 light years across the plane of the galaxy, through a dense swirling mass of stars, gas and dust to the core. The light from the center is completely obscured by interstellar dust, but there are two types of radiation, radio and infrared, that can penetrate the screen.

The pioneering radio astronomer Karl Jansky first recorded a steady hiss coming from the direction of the galactic core, behind the constellation of Sagittarius in the southern part of the Milky Way, as long ago as the 1930s. Radio telescopes have since mapped the center in some detail, but they lack the resolution of infrared images.

Infrared techniques were slow to develop but they were taken up by astronomers in the early 1970s. This image, made by Eric E. Becklin and Gerry Neugebauer in 1979, using the one-meter telescope at Las Campanas Observatory in Chile, shows how sophisticated they have become.

The computer-enhanced image was made at a particular wavelength of infrared (2.2 microns, or about four times the wavelength of green light) emitted by large cool stars called red giants. As these are uniformly distributed, the strength of the radiation gives a good indication of the density of stars in any area. The color code shows the radiation increasing from blue to red. To get this degree of definition, the telescope was used as a scanner, tracking backwards and forwards across an area of sky about the size of two full moons, while the signals were electronically amplified. The thin white line represents the galactic plane and the bright spots are red giants in the foreground. The white cross marks the position of the central core, an extraordinary cluster of stars about three light years across which is spinning at right angles to the galaxy and some 20,000 times faster. Within it dust and debris are bathed in the radiation of 2 million stars, which are packed so closely (less than a light week apart) that they must frequently collide.

There are two pieces of evidence that indicate the presence of a black hole in the core. The first is that the rate of spin suggests a mass of nearly 8 million stars – four times the number that are visible in infrared. The other clue comes from radio observations. In addition to the X-rays they emit (see page 22), black holes should have a characteristic radio shape, like a bright disc, produced by gas and dust as they are sucked across the "event horizon." Radio maps now show a peak like this, which exactly coincides with the center of the infrared image.

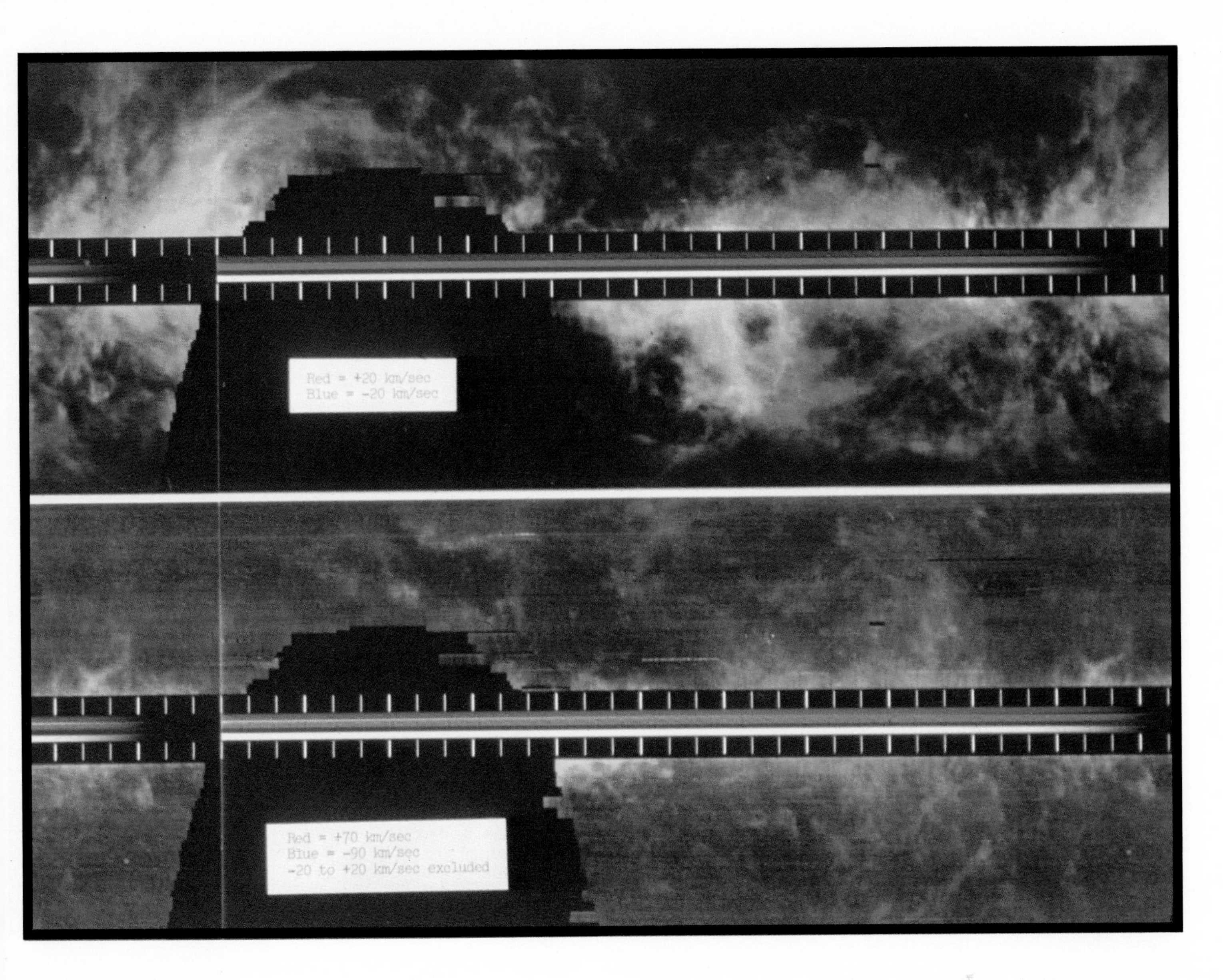

UNIVERSAL GAS

Between the stars and nebulae, in what we think of as the empty vacuum of space, lies a form of matter so thin that it barely exists. The atoms of gas are centimeters apart and you would be lucky to find a single grain of dust in 100,000 cubic meters. But the total volume is so vast that they form well-defined clouds.

In 1978 electronic techniques enabled Dr. Carl Heiles to produce this magnificent 360° panorama showing for the first time the structure of the interstellar medium as seen from our corner of the galaxy.

The clouds are so attenuated as to be almost invisible, but there is one clue to their existence. Although the isolated atoms may collide with each other only once in four hundred years, the activity is enough to produce faint radio waves, familiar to astronomers as a background noise that seems to come from every direction in space.

Dr. Heiles and his colleagues at the Hat Creek Observatory of the University of California carefully scanned the sky with their radio telescope, listening to this noise – the hiss of atomic hydrogen on the 21-centimeter wave band – and recording minute fluctuations in the signals. The measurements were compiled in a computer, and this is the picture that finally emerged.

A broad horizontal band along the plane of the galaxy has been left out because the gas there is so dense that details are obscured. The dark parabola (left) represents the area of the southern sky not visible from California, and the small black rectangle (upper right) is a blind spot where the telescope could not point because of details in its design.

One remarkable discovery was how fast the clouds were moving. The relative brightness shows how much gas there is in any region and the color code indicates the speed and direction in which it is traveling. Gas receding from us at up to 20 kilometers a second is colored red; that approaching us at similar speed is blue; and material that is stationary (relative to the solar system) is yellow.

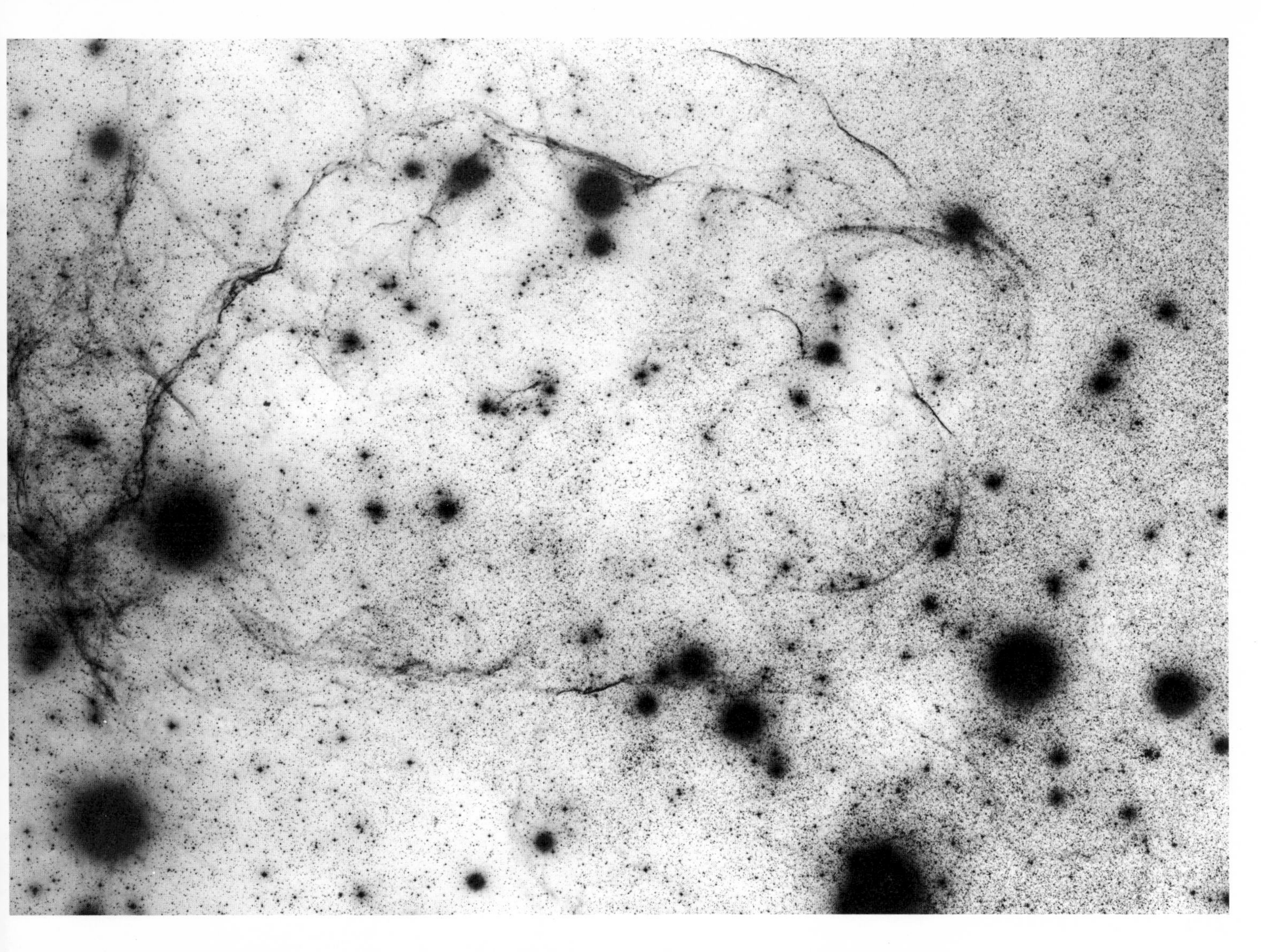

ABOVE: **The shockwaves of a supernova spread out through space. The explosions of old stars enrich the galaxy by distributing the material they have manufactured and triggering off the birth of new ones.**

This supernova occurred about a millennium ago, 1000 light years away in the constellation of Vela. Similar explosions occur somewhere in our galaxy every century or so, but only four have been close enough to be seen from the earth since records were first kept.

OPPOSITE: **A group of young stars signal their arrival by lighting up the beautiful loops and filaments of the Orion Nebula. The stars, the Trapezium Cluster in the region of Orion's sword, produce this effect by ionizing hydrogen atoms in the gas around them, which causes it to fluoresce.**

The nebula is part of a much larger cloud in which still younger stars are developing. These are not yet visible but are giving off strong infrared radiation from behind, and immediately above, the nebula.

The feature on the left side, called the Dark Bay, is not a nick out of the edge but simply part of the cloud which has not yet been ionized.

THE BIRTH AND DEATH OF STARS

Stars form inside clouds like the Orion Nebula from a cold, dense mixture of molecular hydrogen (H_2), carbon monoxide (CO), grains of silicate and carbon dust, and a sprinkling of other elements. But the main ingredient is gravity. It is gravity alone which shapes every event in a star's life and which transmutes these unpromising materials quite literally into gold.

The individual atoms in a cloud of gas are usually far enough apart not to be affected by it, but if they are compressed by a shock wave, some pockets of gas may become dense enough to form their own weak gravitational fields. As it pulls away from the rest of the cloud, the field intensifies and the gas begins to respond to a remorseless and steadily increasing attraction toward its own center. The initial contraction can take up to 10 million years, but eventually the core shrinks to a liquid nucleus, dragging the rest of the gas in after it.

As the material gets denser it begins to heat up, and when it reaches 30°K, the object starts to emit infrared radiation, the first sign of its existence.

The radiation slows down the contraction, but it continues to shrink until it reaches the critical temperature of 10^7K and the nuclear furnace ignites. This halts the contraction (for a while at least) and the star begins to shine.

As long as it is turning hydrogen into helium, enough energy is released to counteract the gravity. Big stars burn quicker than small ones, but sooner or later the fuel always runs out and the star starts to shrink again.

The pressure and temperature steadily rise until conditions are suitable for another nuclear reaction to begin, turning the helium into heavier elements, like carbon. And when the helium runs out, the screw tightens still further, until the carbon can be used as a fuel, by being turned into even heavier elements like iron.

As the nuclear chemistry gets more complicated, the star becomes increasingly unstable, and each collapse is more violent than the last. Eventually the core implodes with such force that the outer layers of the star are blown off in a massive blast.

The largest of these explosions, a supernova, is the most violent process in a star's life and, paradoxically, it is the only one with enough energy to produce the really heavy elements. All the uranium on earth, all the gold and silver and platinum, all the mercury, and even the iodine, was formed in these brief moments of a dying star.

Depending on its size, the core becomes a white dwarf or one of the ultradense objects that fascinate modern astronomers today, such as pulsars and neutron stars. If it is really big and the gravity is strong enough, it becomes a black hole.

The flash of a supernova can be brighter than 10 million suns, and the fireball expands at over 10,000 miles a second for about a century, until the interstellar medium slows it down and the gases are compressed into a shell.

The shockwaves continue to spread out through the galaxy for hundreds of thousands of years, depositing their load of gold and iodine in the form of dust and triggering off the birth of other stars. Gradually slowing down and cooling, they eventually lose their identity in the interstellar medium.

OPPOSITE: **The Carinae Nebula, one of the finest sights in the southern sky, contains a strange star that appears to be dying and giving birth at the same time.**

Eta Carinae flared up during the nineteenth century and by 1846 it was the brightest star in the sky after Sirius. Then it faded, apparently without exploding, and has sputtered off and on several times since, though it remains the strongest source of infrared radiation in the Milky Way.

It could be a supernova muffled in dense clouds or a star shedding its material in a series of partial collapses but it is certainly providing chain reactions within the nebula. Recent studies by the orbiting X-ray telescope of the Einstein Satellite show that the diffuse source at the center is surrounded by seven point-like emissions of X-rays, indicating the birth of new stars.

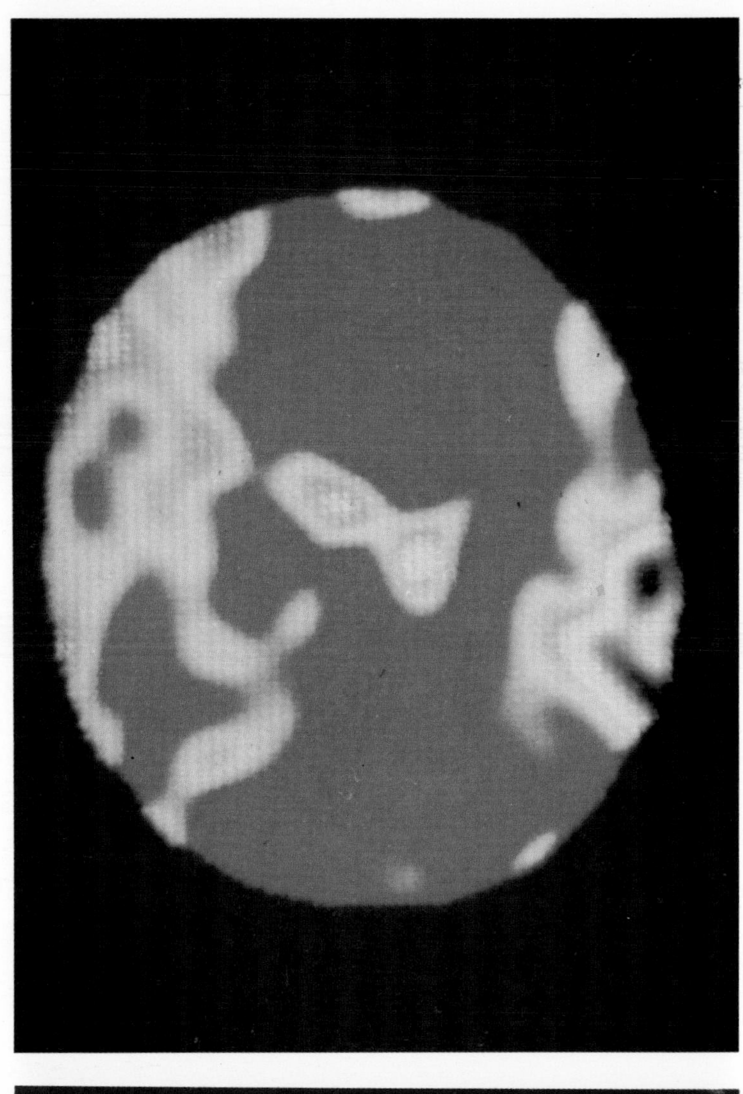

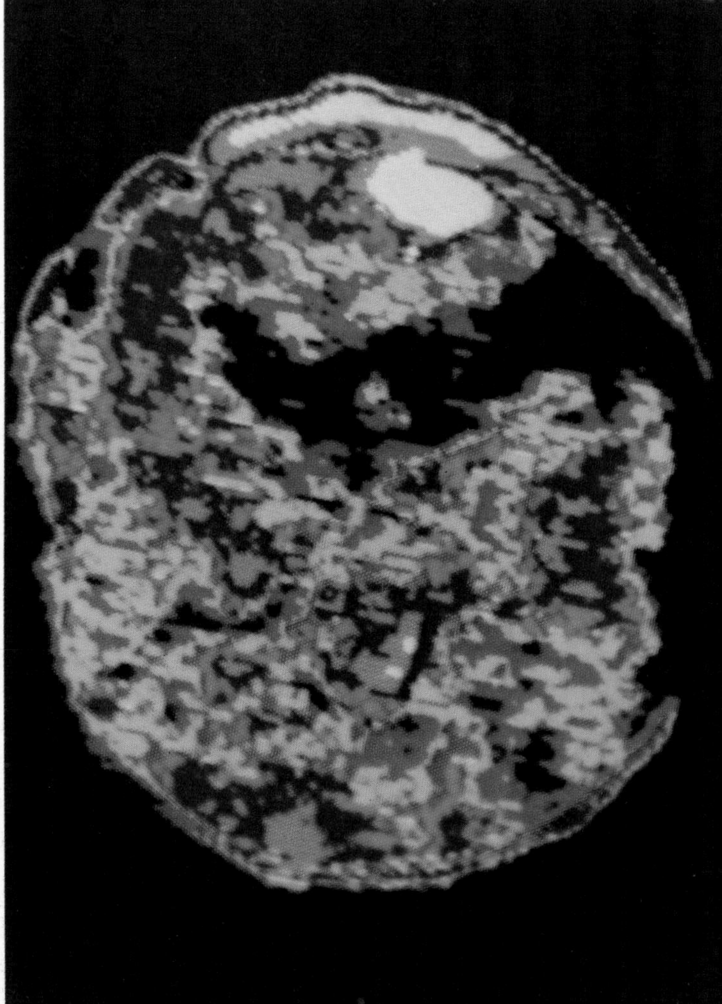

THE SURFACE OF A STAR

It is an odd fact that, although the stars have been studied for centuries, we still have no idea what they actually look like. The pinpricks of light are too small to have any discernible shape, and in the whole universe only two of them are close enough for us to make out the features on their surface. One is Betelgeuse, 650 light years away in the constellation of Orion (TOP LEFT). The other, 8½ light minutes away, is the sun (BOTTOM LEFT).

The sun, of course, has been photographed many times – this is a computer-enhanced X-ray image of it – but the picture of Betelgeuse is unique.

There is no point in enlarging a photograph of a star to see what it looks like, because the atmospheric distortion that makes it twinkle produces a blurred image on the emulsion. However, very short exposures of up to 1/100th of a second show that the light is made up of dots or speckles, which can now be analyzed by a technique called speckle interferometry to reveal structural information about the source.

The techniques are normally only used to measure the diameters of stars, but by comparing a number of these exposures the computer at the Kitt Peak Observatory was able to build up this image of the surface.

Betelgeuse is 800 times the size of the sun and, though interpretation is difficult, it appears to be roughly "faceted," with fewer and larger cells on its surface.

THE SHELLS OF BETELGEUSE (RIGHT)

With the discovery of the solar "wind" about twenty years ago, it was realized that visible stars were only the core of immense systems in space, and astronomers began looking for similar structures around our neighbors.

The picture opposite, an artificially colored contour map of potassium light intensities, shows the clouds of gas that were found to be billowing from the surface of Betelgeuse.

The outer shells of gas are many hundreds of times larger than the star's own radius, and at the edge of the system, where the gravity is weak, the gas streams off into space.

The events taking place around stars are still largely unknown. For instance, the gas not only leaks away but seems to be forcibly ejected by some unknown mechanism. Since these processes could drastically affect the life span of a star, including our own sun, they are now a major target of research.

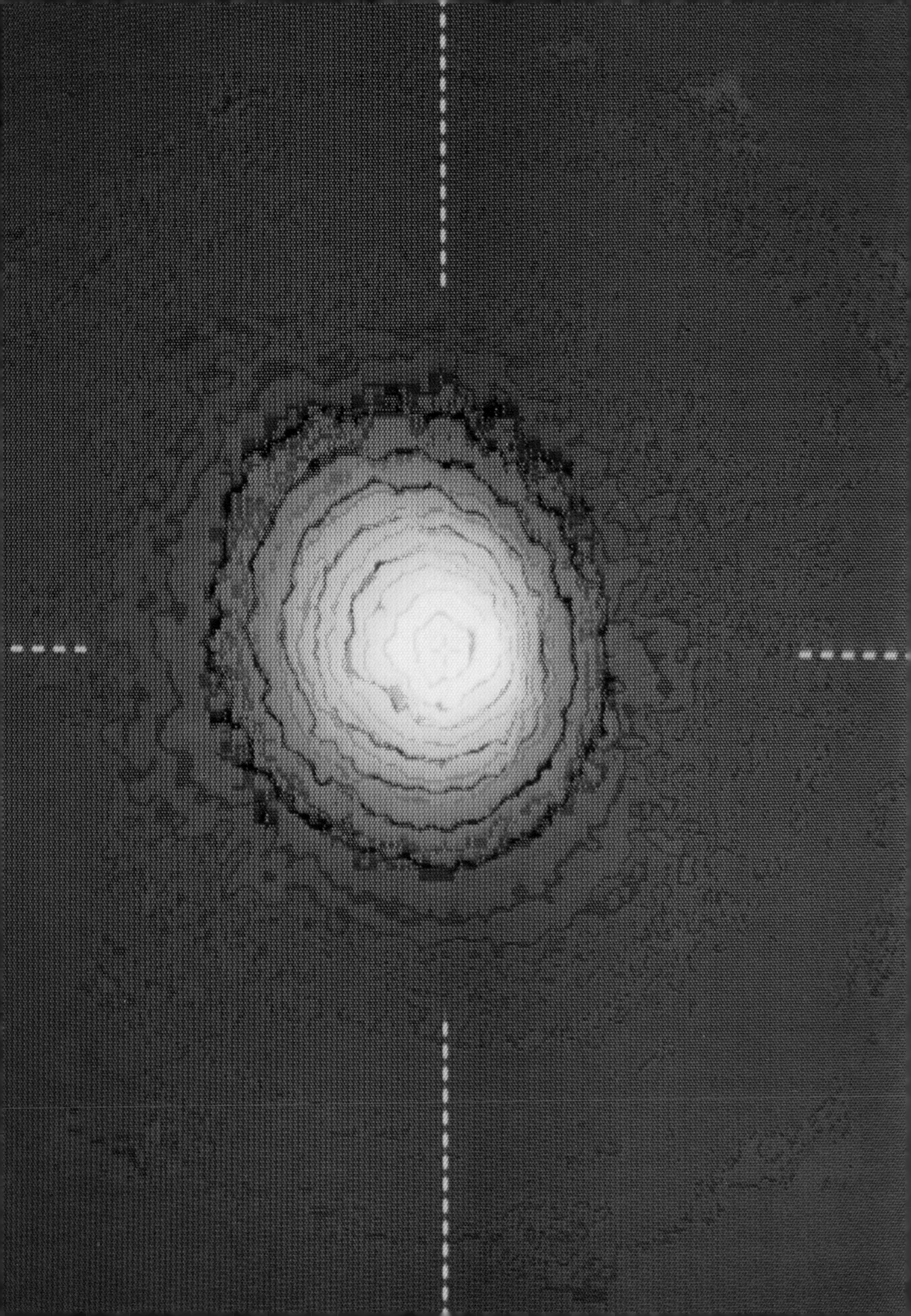

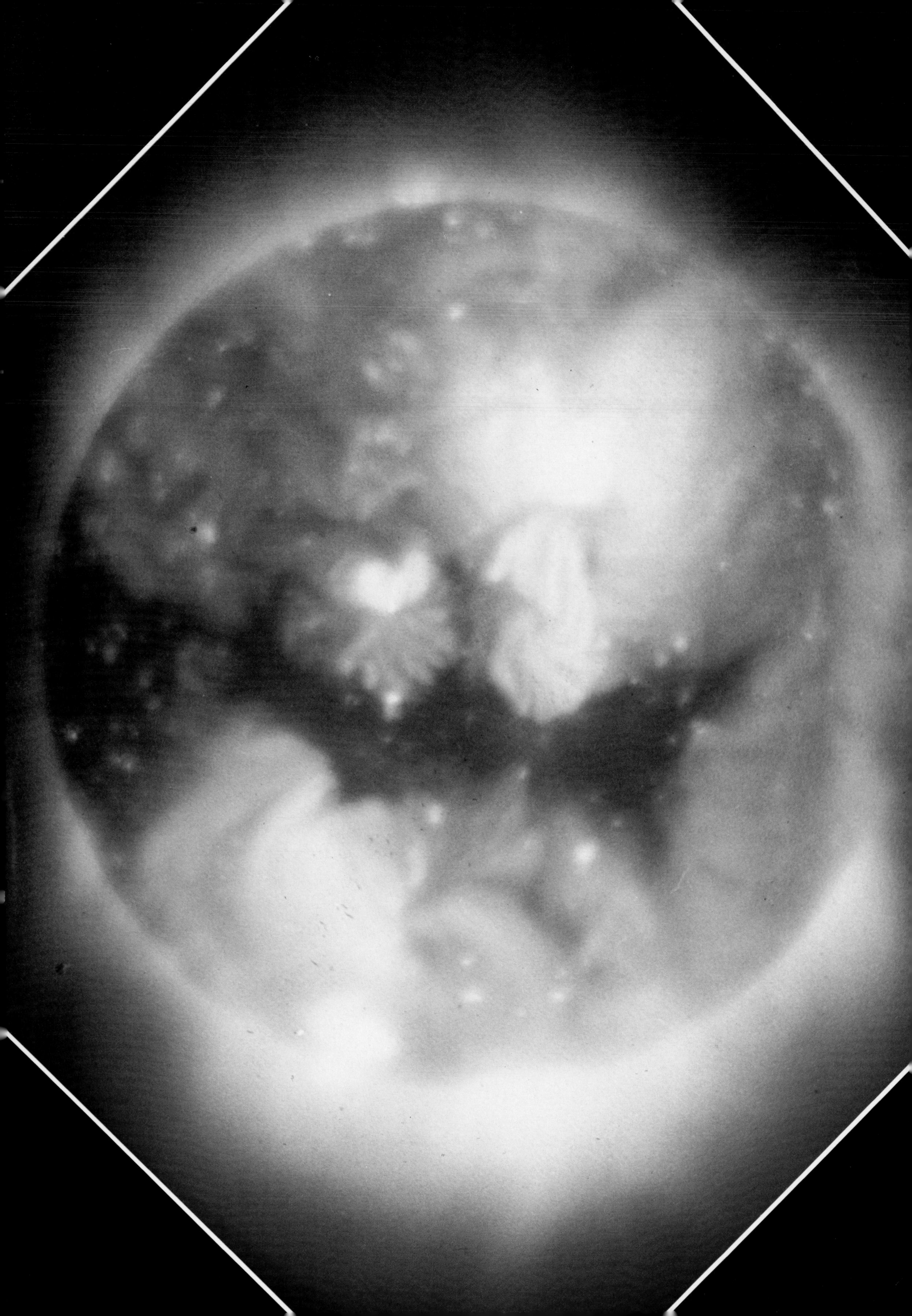

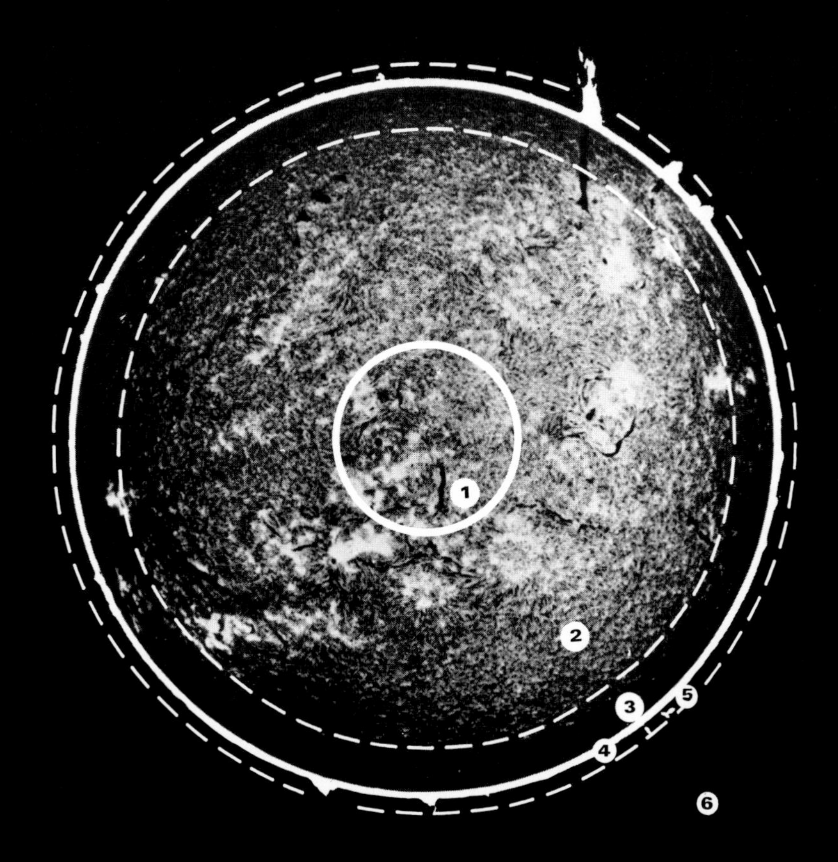

The structure of the sun, superimposed on a picture of the photosphere, the skin of bright, cool gas from which we receive sunlight. The surface, which is no more than 300 miles (500km) thick, was photographed in the light of ionized hydrogen.
The diagram shows the relative size of (1) the nuclear explosion in the core; (2) the dense, solid plasma around it; (3) liquid currents below the surface; (4) bright surface gas (the photosphere); (5) the solar atmosphere (the chromosphere); (6) thin outer gases (the corona)

THE SUN

**"Doubt thou that the stars are fire?
Doubt that the sun doth move?"**
– *William Shakespeare*

SKYLAB

The first orbiting observatory not only opened a new era of solar astronomy, but had the unusual distinction of being destroyed by one of its own discoveries.

The blanket of the earth's atmosphere both protects us from and blinds us to most of what is happening in space. Apart from the narrow window of visible light, the only types of radiation that get through are very high frequencies (such as cosmic rays), very low frequencies (like radio) and a few precise wavelengths of infrared. The rest of the universe is literally invisible to us.

Experiments with rockets and balloons enabled us to catch a brief glimpse of what lay beyond, but it was only with the launch of Skylab in 1973 that we were able to take a long cool look at the universe for the first time. The spacecraft carried a mass of equipment including two X-ray telescopes, ultraviolet cameras, spectrometers and radiometers of various kinds and a huge multi-spectral unit with no less than thirteen detectors, including thermal infrared. Only five of the thirty cameras on board were shooting in visible light; the bulk of the instruments were designed to record the invisible frequencies. Some of them were pointed down at the

LEFT: **An X-ray of the sun, taken by the crew of the US orbital spacecraft Skylab. In the light of X-rays it is possible to see huge chasms reaching to the core, which are venting gas into space.**

ABOVE: **The bright, cool surface of the sun. A photograph taken in the light of ionized hydrogen of the photosphere.**

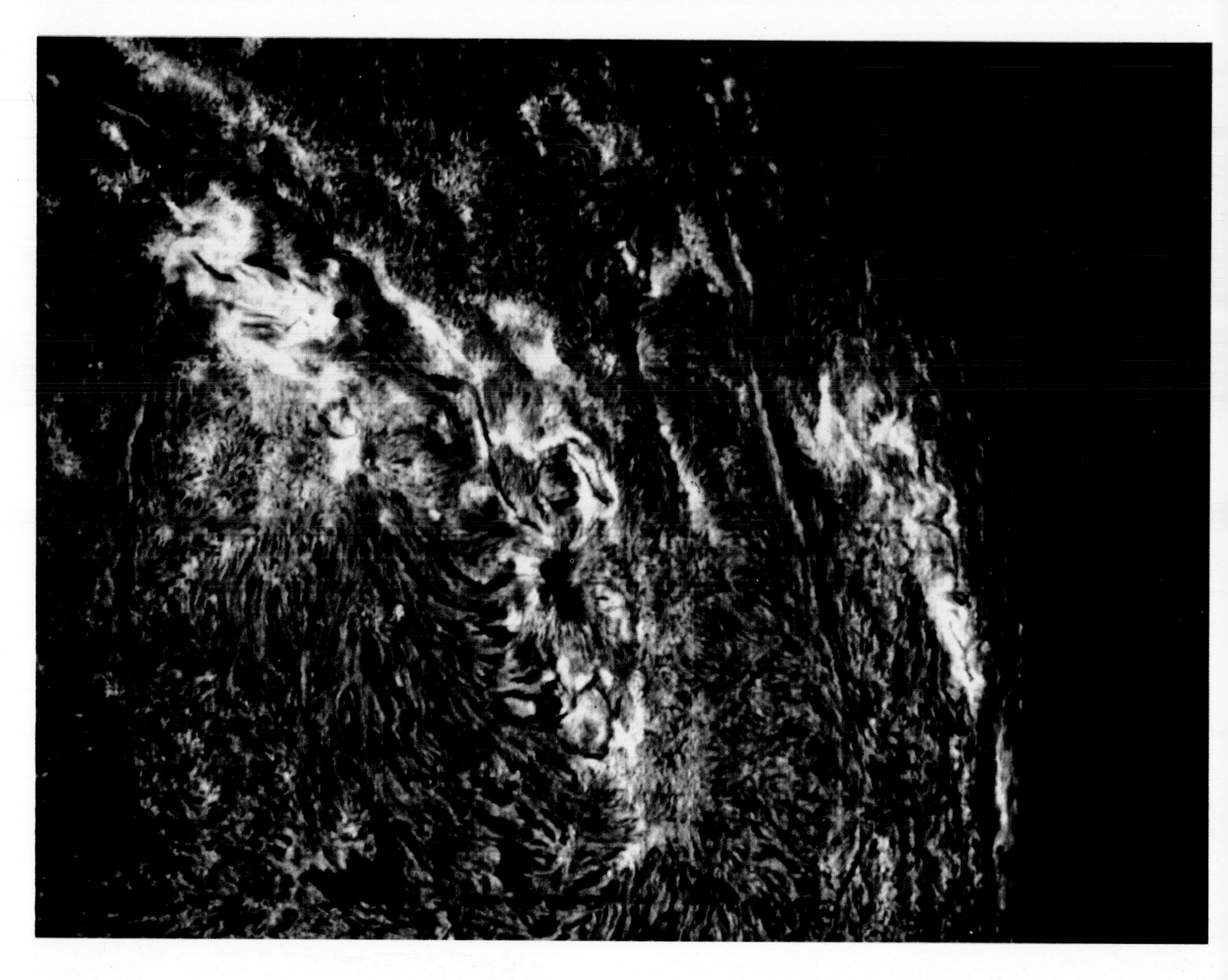

ABOVE: **The frothing, bubbling surface of the sun, which radiates 70,000 horsepower of energy from every square meter – the equivalent of burning 11 million times the world's annual coal output every second. Yet, paradoxically, the photosphere is one of the coolest parts of the sun. The atmosphere, or chromosphere, above it is a million degrees hotter.**

earth and examples of their imagery are reproduced in the final section. Skylab's primary mission, however, was to study phenomena in space; and the main target was the sun itself.

The discoveries made by the three shifts of astronauts who manned the spacecraft altered many previous assumptions. A detailed study of the surface, using spectroheliometers and two spectrographs (with nine 35mm cameras), revealed that the giant cellular structure was far more complex than anyone imagined. Patterns of storms were detected across the whole solar disc, while the X-rays showed that the corona was composed of closed loops of gas with "coronal holes" emitting violent belches of solar wind.

Their observations showed that this stream of ionized particles was strong enough to deflect the earth's magnetic field. It was found to affect the earth's atmosphere, compressing it when the solar wind was strong enough and allowing it to expand again when the pressure eased up.

This last discovery was critical to the project. Skylab had been put into orbit just above the atmosphere at a height of 270 miles (432km). NASA had timed the launch for the beginning of a quiet period of solar activity to minimize the radiation hazard to the crew, but they realized, too late, that the combination of factors actually threatened the spacecraft. As the solar wind died down, the atmosphere rose beneath the craft and the drag slowed it down. The orbit shrank as it met more and more resistance until, like a ship in shallow water, it eventually "ran aground" in the summer of 1979. Unable to boost it into a higher orbit, the NASA controllers used what was left of the fuel to control its descent and watched helplessly as 2.6 million dollars' worth of equipment burned out over the Australian outback.

In spite of its premature end, Skylab made a major contribution to solar astronomy, and added to a growing interest in the subject. After all, the sun is the only star that astronomers will be able to analyze in detail for many years to come, and for the first time in centuries of frustrated guesswork and religious bigotry, they have real information to go on.

Now that nuclear physics can explain some of the processes and spacecraft have provided tangible evidence to confirm the theories, we can begin to understand what a remarkable object it is.

THE DARK STAR

On a cosmic scale the sun is an average, rather undistinguished star. Antares is 27 million times larger, Rigel is 50,000 times brighter and, at the other end of the scale, the diminutive Wolf 457 is thought to be smaller than the earth. But in human terms even an average star is awesome.

The sunlight that powers our planet comes from an H-bomb explosion held in place by magnetic fields and the sun's own immense gravity, consuming 14 million tons of fuel a second, for a period of 10 billion years.

The explosion – a proton-proton cycle turning hydrogen into helium – is now about halfway through. It is estimated that the hydrogen in the core has dropped from 75 percent to 35 percent. When it finally runs out, slower and less violent processes will take over. There will no longer be enough outward pressure to withstand the gravity and theoretically the core should collapse into a solid object about the size of the earth, while an expanding cloud of superheated gas wipes out the solar system.

Strangely enough, the energy of sunshine takes literally millions of years to reach us from the center of this solar furnace. Most of the time is taken as the energy inches its way out along sluggish convection currents from the core of compressed plasma, a black soup of atomic fragments twelve times as dense as lead. Halfway out to the surface, the plasma has the consistency of water, but it is still pitch dark because the partly reformed (but still unstable) atoms make an opaque barrier to radiation.

About 95,000 miles (150,000km) below the surface the plasma becomes a gas and by this time the temperature has dropped from 15 million°K in the core to a mere 3000°K. As the gases expand and are carried up through the turbulent storms to the surface, a curious temperature inversion takes place. They begin to heat up again and glow like moonlight; when they reach the surface a few weeks later, they are 2000° hotter and radiating brilliant light.

This bright surface is the *photosphere* from which we receive sunlight, and above it are the thin invisible gases of the *chromosphere*. Though they are so attenuated that they emit very little light, they continue to expand and get hotter, reaching 2 million°K as they billow out into the *corona* and beyond that into space as the million-mile-an-hour solar wind.

The energy that took so long to reach the photosphere completes its journey to earth more rapidly. The radiation reaches us in 8½ minutes at the speed of light. The first high-speed particles arrive on a tight loop through space about an hour later, and a gust of solar wind makes the journey in two days.

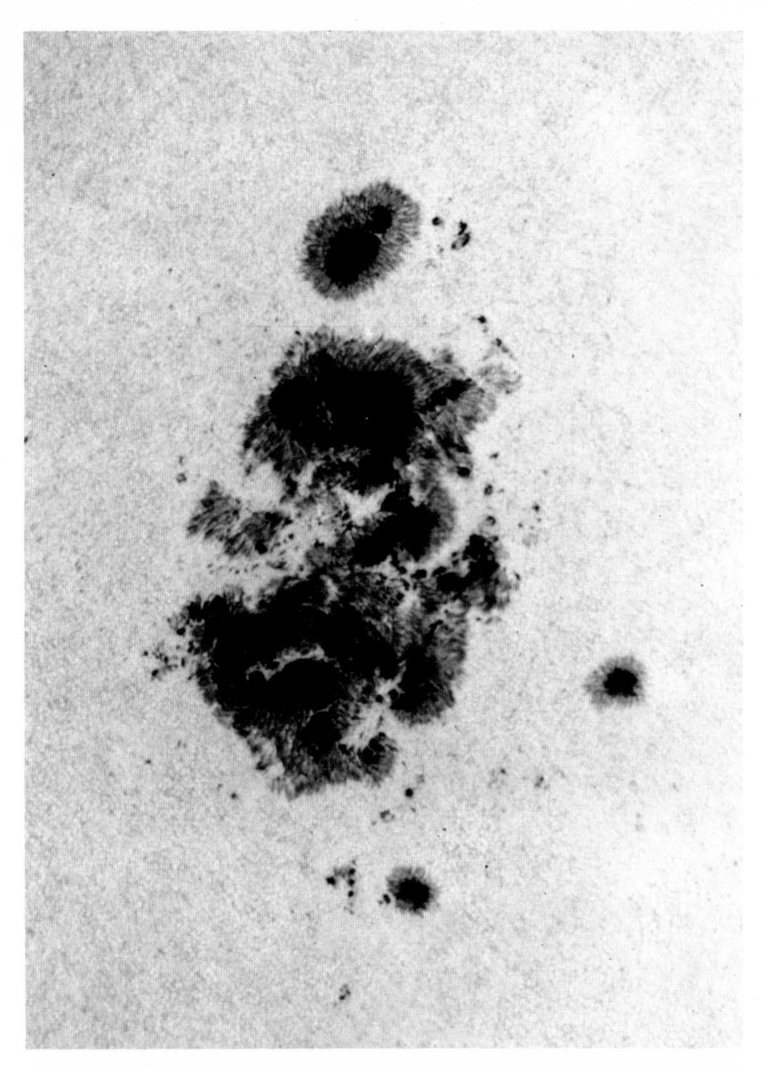

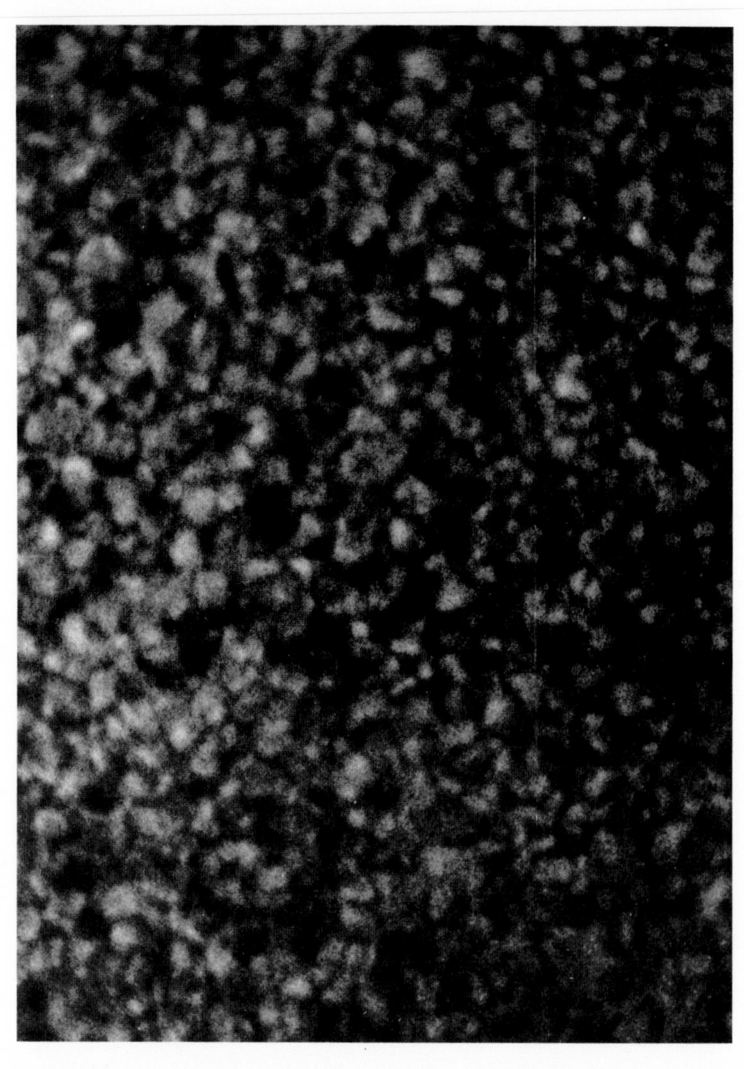

ABOVE RIGHT: **A group of sunspots, several times the size of earth, drifts towards the solar equator.**

BOTTOM RIGHT: **The grainy surface of the sun's photosphere. The strange texture is caused by geysers of gas up to 625 miles (1,000km) across, welling up in the center of each "grain" and falling back around the sides.**

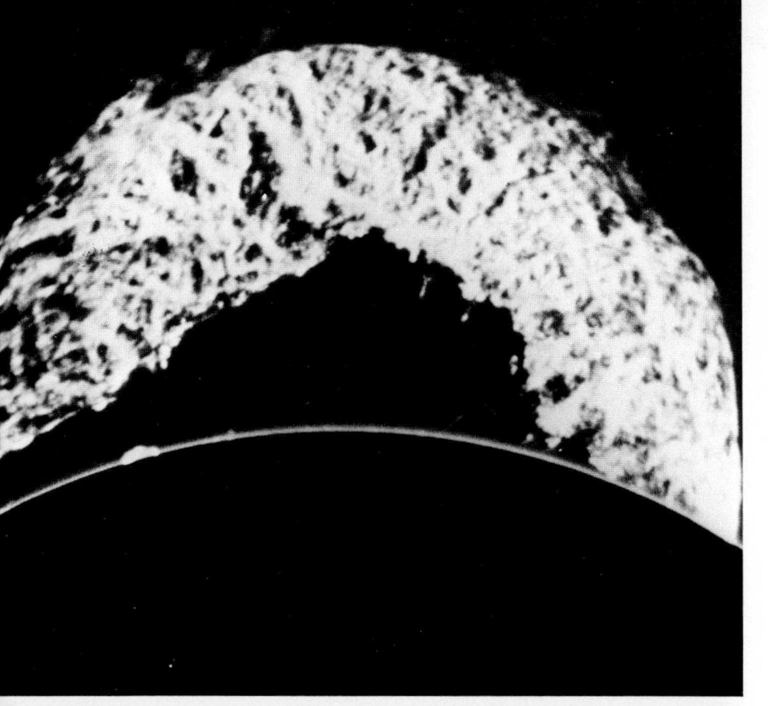

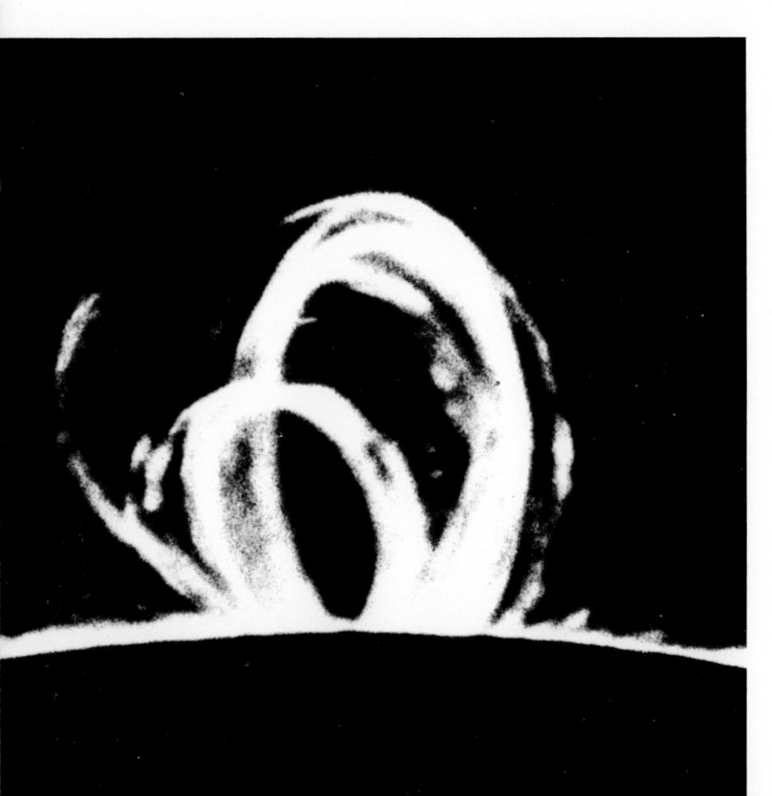

SUNSPOTS AND CYCLES

The surface of the photosphere is an unstable skin, and the magnetic network that holds it together is easily disturbed.

Sunspots occur where a knot of intense magnetism breaks up the surface and reveals the darker gases underneath, which look black against the brightness. Rather like meteorological depressions in the earth's atmosphere, sunspots drift across the surface of the sun and often develop in pairs. Reactions between them can lead to the complete collapse of surface magnetism, allowing the stronger magnetic lines inside the sun to balloon out, like a massive hernia, into space.

The shapes of these magnetic loops are lit up by the gas flowing along them, and their size can be staggering, with literally billions of tons of gas hurled through space at speeds of up to 300 miles (500km) per second for hundreds of thousands of miles.

Sunspots and other pyrotechnic displays are known to come and go in cycles of roughly twenty-two years, which appear to affect the earth's weather patterns. For instance, the most intense period of the "Little Ice Age," between 1645 and 1715, exactly coincides with the quiet period of solar activity known as the Maunder Minimum. The sunspot cycle has even been associated with such terrestrial phenomena as the fluctuation of the price of corn on the New York Stock Exchange and the incidence of auto accidents.

Buffeted as we are by the solar wind and bombarded by both particles and radiation, it is not surprising that we are affected by solar events. The first tangible evidence of this appeared in 1972, when the shockwave from a solar flare was tracked from the sun to earth by an orbiting US satellite. Traveling at 125 miles (200km) a second, the shock's arrival coincided with the beginning of a storm in the earth's ionosphere that caused auroral displays and wreaked havoc with short-wave radio communications.

TOP: **Gas coils along a spiraling magnetic field, getting brighter and hotter as it expands.**

CENTER: **Appearances can be deceptive. These are not flames rising, but gas falling back to the surface down a relatively stable fringe of magnetic tubes around each "grain."**

BOTTOM: **Complex loops of magnetism drag material out from the surface and back again.**

OPPOSITE: **Skylab cameras caught this gigantic flare arching 360,000 miles (576,000km) over the sun's surface, lifting as much as a thousandth of the surface gases into space at one time.**

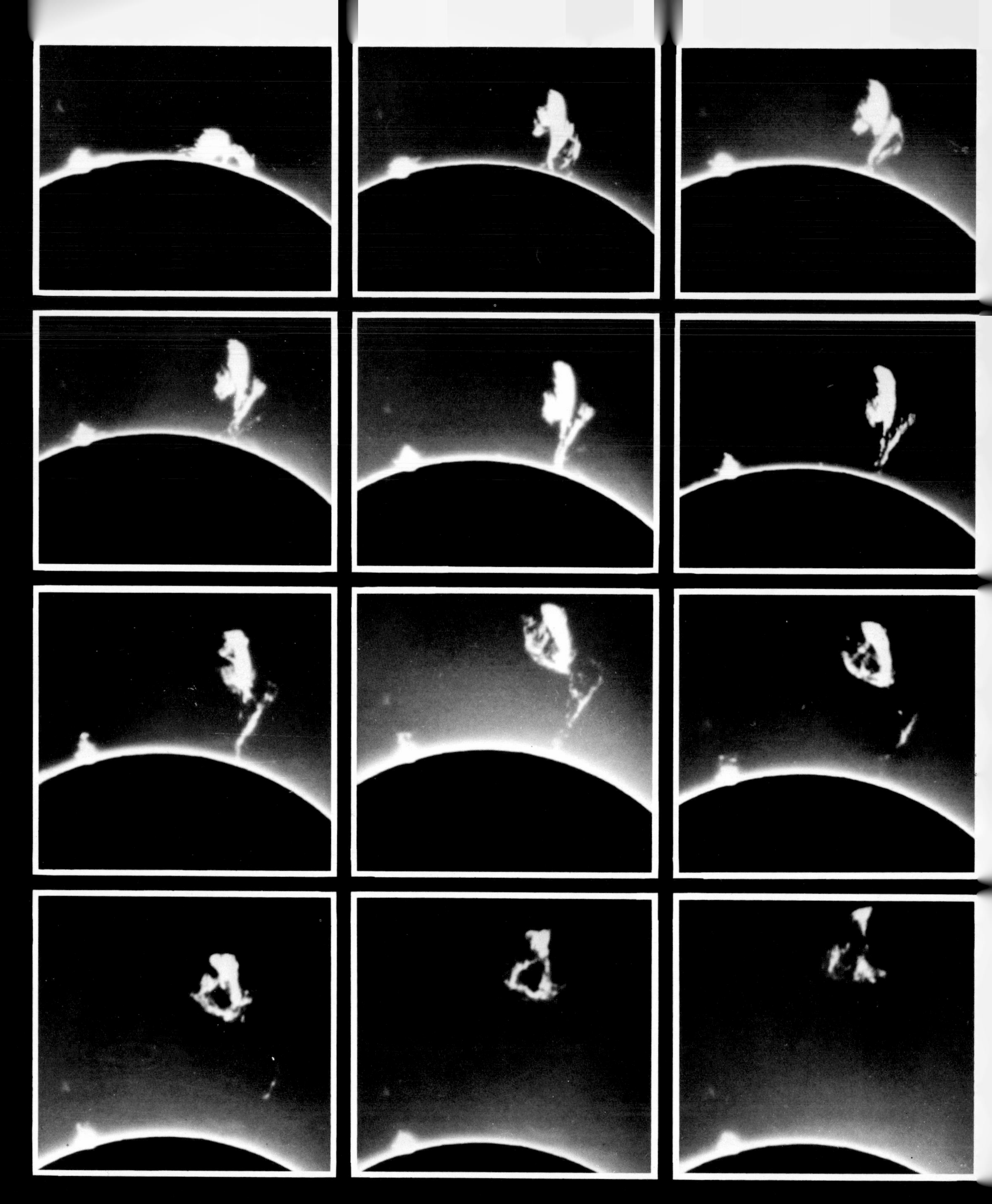

Stop-frame photography shows the escape of a solar flare as it breaks free of its magnetic bonds. The region of the sun shown here is roughly 350,000 miles (560,000km) across and the event lasted about 30 minutes, so the flare was moving at over a quarter of a million miles an hour.

Flares look like tongues of fire, but they are the exact opposite. These pictures are negative images, where the usual association between brightness and heat is reversed; the "flames" are actually eruptions of cold gas into the hot darkness of space surrounding the star.

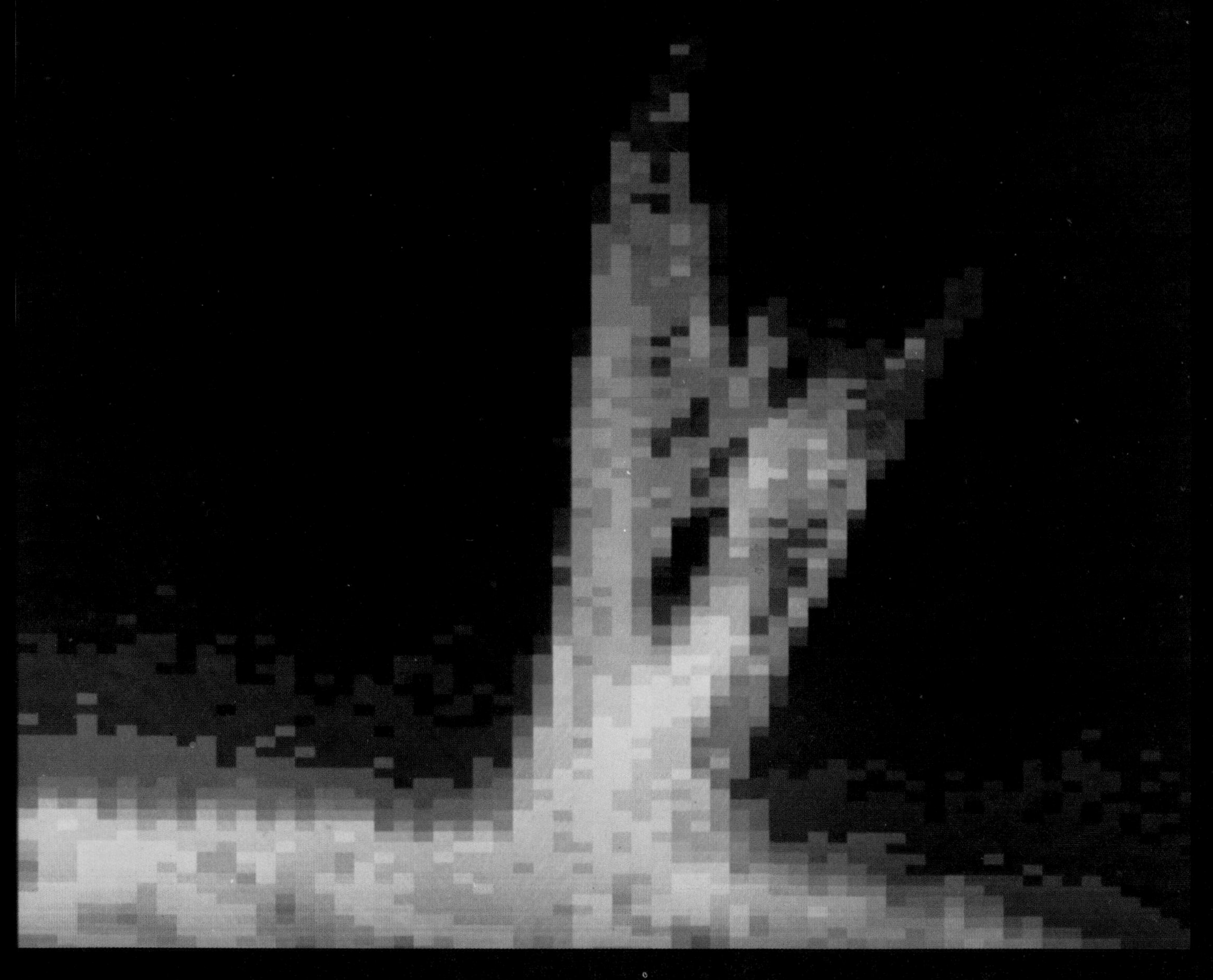

An ultraviolet study of a solar flare, made from Skylab. The spectrometer recorded two frequencies simultaneously to show the transition from the gas of the flare, at 10,000°K, to that of the corona, at 2 million°K around it. The darker the color, the higher the temperature.

The occasional squares of green represent the light from neutral hydrogen inside the flare. The red squares that dominate the picture represent radiation from highly ionized oxygen on the surface, where the atoms are being stripped of their electrons by the heat.

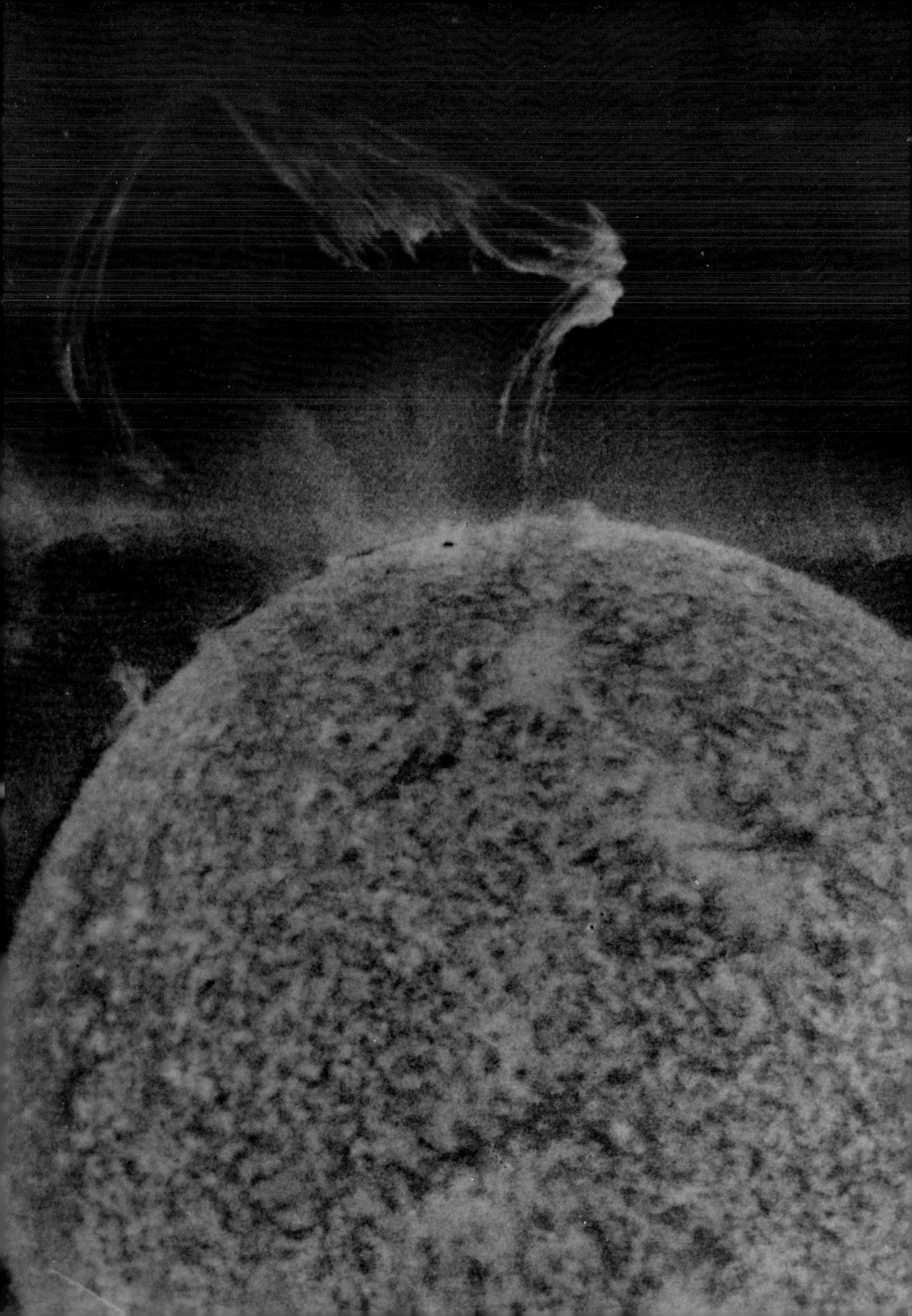

THE GIANT FLARES

This huge solar flare (LEFT), reaching more than 600,000 miles (1,000,000 km) into space, was one of the most spectacular events witnessed by the crew of Skylab.

The loops and fringes hovering in the sun's atmosphere seldom rise more than 30,000 miles (48,000km) above the surface and are often stable enough to last for days. Giant flares like this, which occasionally burst from the center of sunspots, are an order of magnitude larger and happen at very high speed.

The cloud of ionized helium looks graceful, but it was accompanied by a blast of ultraviolet radiation and a shock wave that tore through the chromosphere. In less than a minute X-ray emissions from the sun increased 100 times; and particles such as protons and electrons were ejected so fast that radio telescopes detected flashes of synchrotron radiation. The gas appears to have run into an invisible barrier in space, but this can be explained by comparing it to the picture of the corona (OVERLEAF), which shows that it corresponds to the outer limits of the magnetic field.

On another occasion the Skylab astronauts were astonished to see a gas cloud, described as "a great transient blob" the size of the sun itself, leave the main body like a ghost and slip out through the chromosphere at 250 miles (400km) a second.

ABOVE: **Computer enhancement of the flare shows that it was part of an even larger eruption. The color code represents various densities of helium gas: the darker the color, the thicker the gas. It is difficult to see the whole of gaseous structures like this because different parts of the cloud are visible at different frequencies, according to their temperature and density. The problem is even more acute when it comes to producing an image of the thin outer gases in the corona.**

OVERLEAF: **The full glory of the sun's corona. With the solar disc masked, a Skylab photograph reveals the layers of thin gas trapped around the sun and the magnetic fields that hold them in place. The corona can only be seen from earth during brief periods of an eclipse, as a faint haze extending out for about half the diameter of the sun. But it stands out clearly against the black background of space, and during the first two months of the Skylab operation it was studied for longer, and in more detail, than had previously been possible in the whole of human history. Eight and a half months of solar observation were eventually recorded, compared to the 80½ hours of natural eclipses recorded since the advent of photography in 1839.**

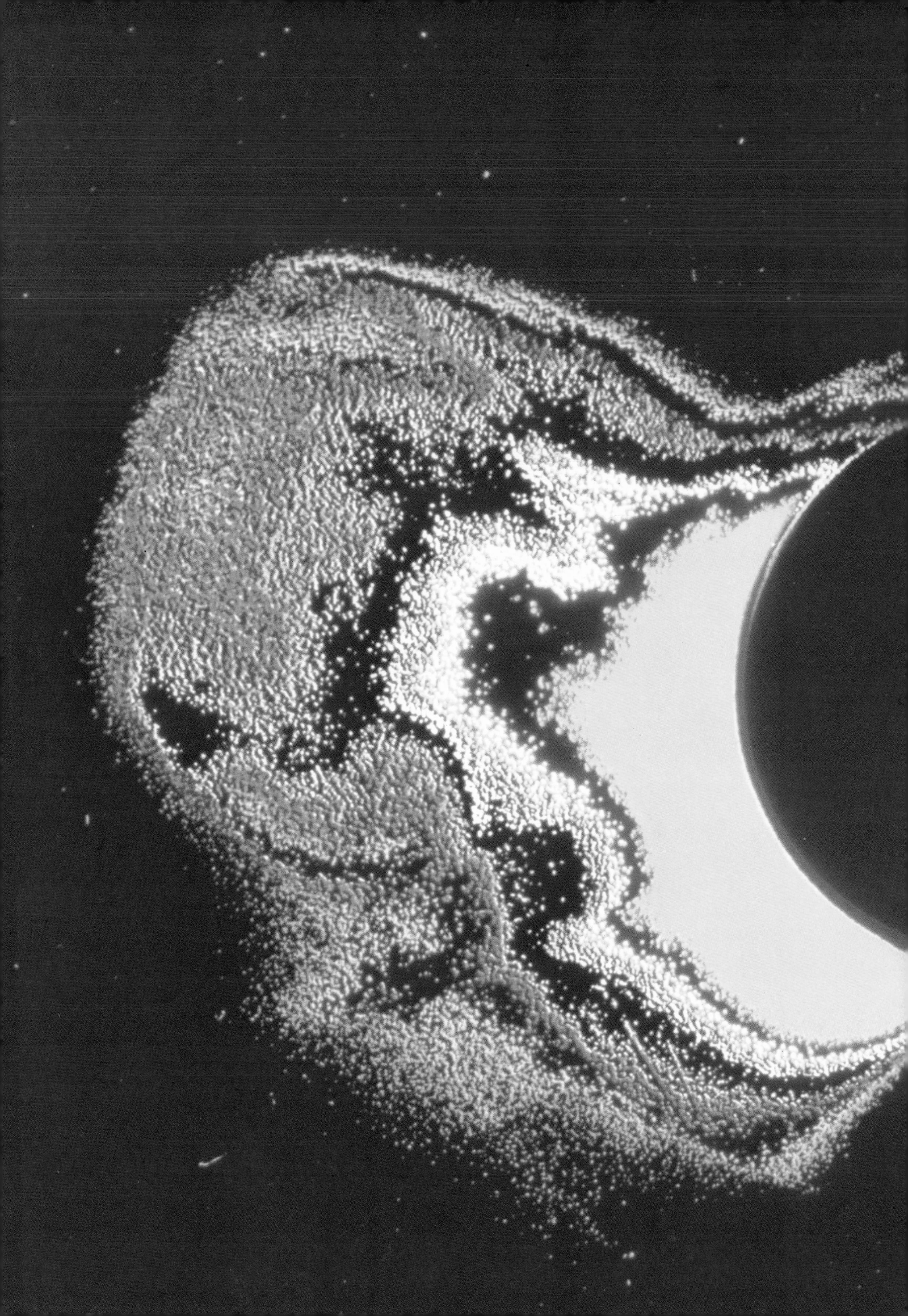

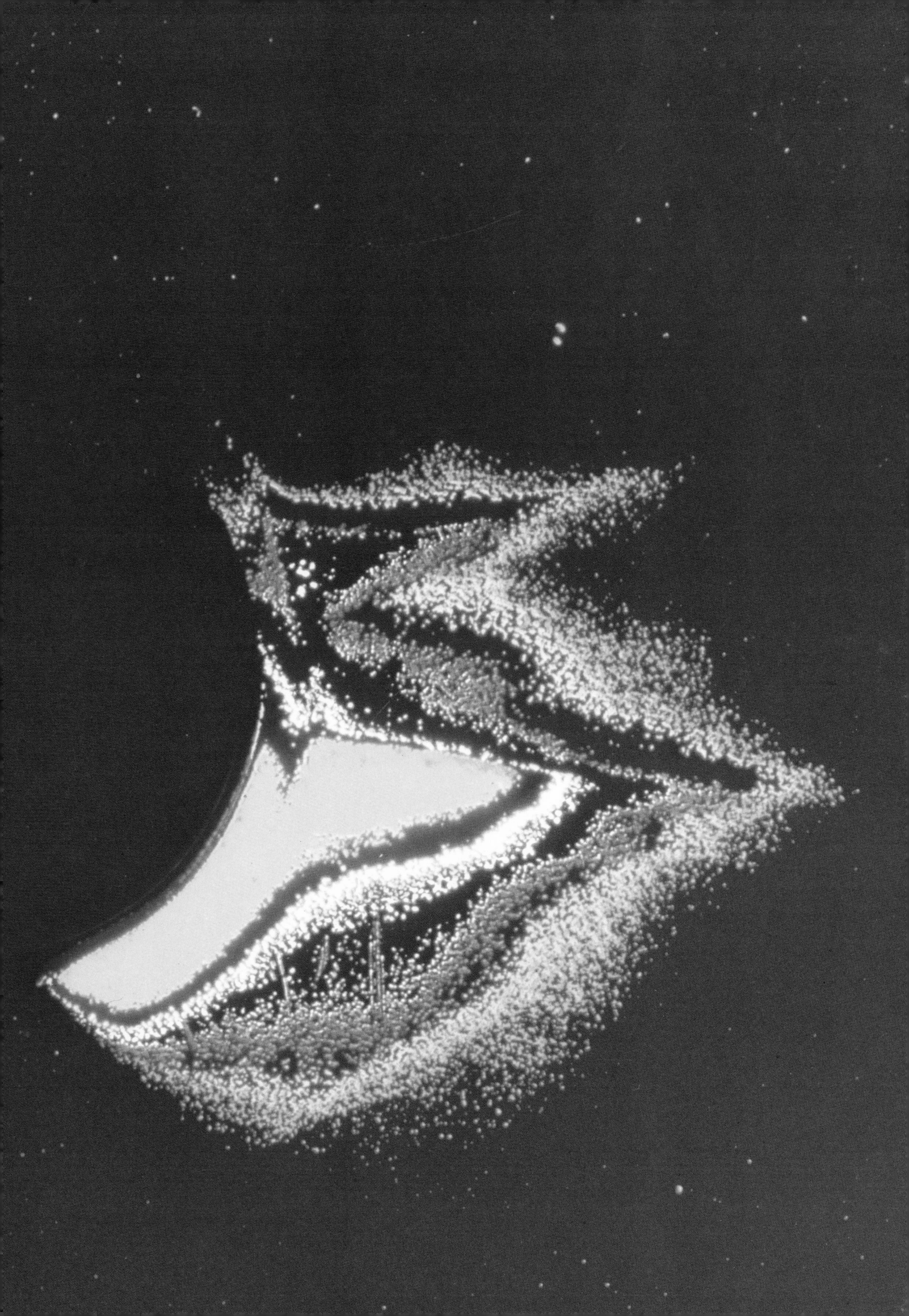

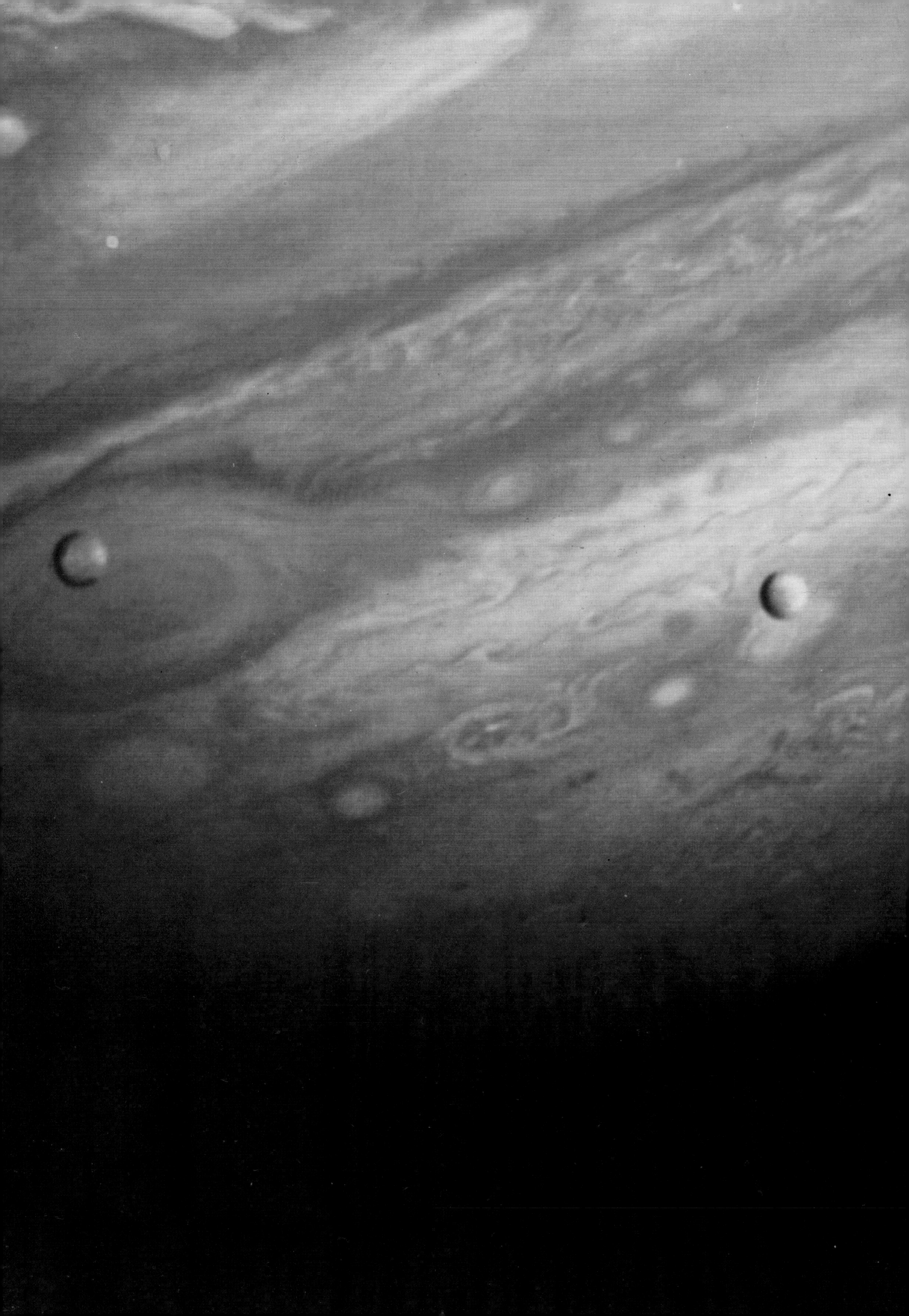

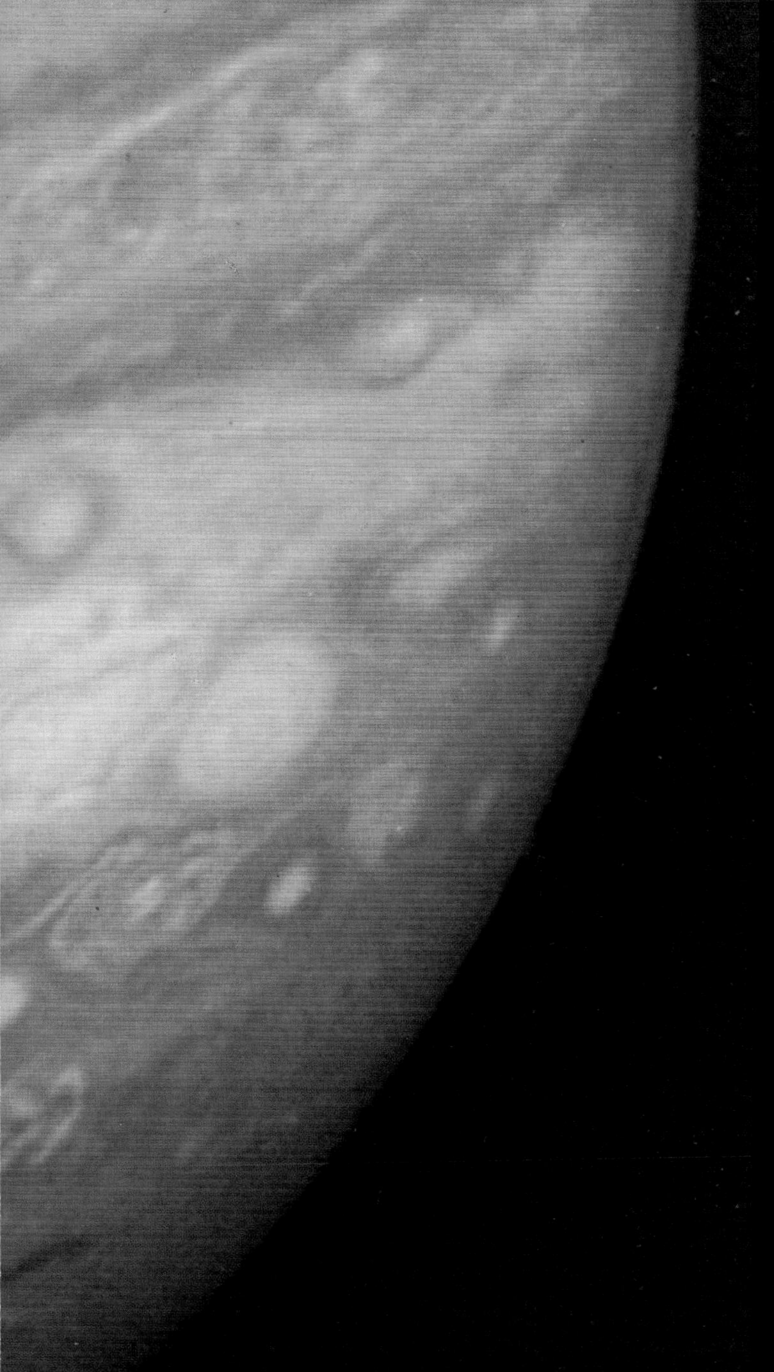

THE PLANETS

"The earth is the cradle of the mind, but you cannot live in the cradle forever."

– Constantin Tsiolkovskii

Tsiolkovskii was a self-educated inventor of Polish origins who forecast most of our present space technology, from solar batteries to the effects of zero gravity, in the early years of the century.

Honored in Russia as the "father of modern rocketry," Tsiolkovskii was a visionary who saw the human colonization of the planets as natural and inevitable. Over thousands of years he envisaged the solar system transformed to support a population 10^{14} times that of earth – and even that would only be the beginning.

"Is it possible for one island to be inhabited," he asked, "and for other islands to be uninhabited?"

We are already ahead of his schedule. We have run our hands through moon sand and watched the sunset on Mars. Our robots, like tiny electronic insects, have circled the planets and rendezvoused with asteroids. Their sensors have tasted the acrid smog on Venus and the reek of garlic in Jupiter's cloudscape; and through their eyes we have seen the lurid sulphur fields of Io and the ice ocean of Callisto, frozen into mile-high tidal waves.

Strange landscapes, clouds denser than chewing gum and million-mile-an-hour winds – in the last twenty years the solar system has proved more varied, colorful and extraordinary than even Tsiolkovskii could have imagined.

As the Voyager spacecraft swung through the radiation belts of Jupiter it recorded two of the moons scudding across the face of the gaseous giant. Jupiter contains twice the mass of the other planets put together. If it had been much larger, the solar system (like many other star systems) would have had two suns. As it is, the planet radiates two and a half times the amount of energy it absorbs from the sun and is bright enough to cast a shadow on earth.

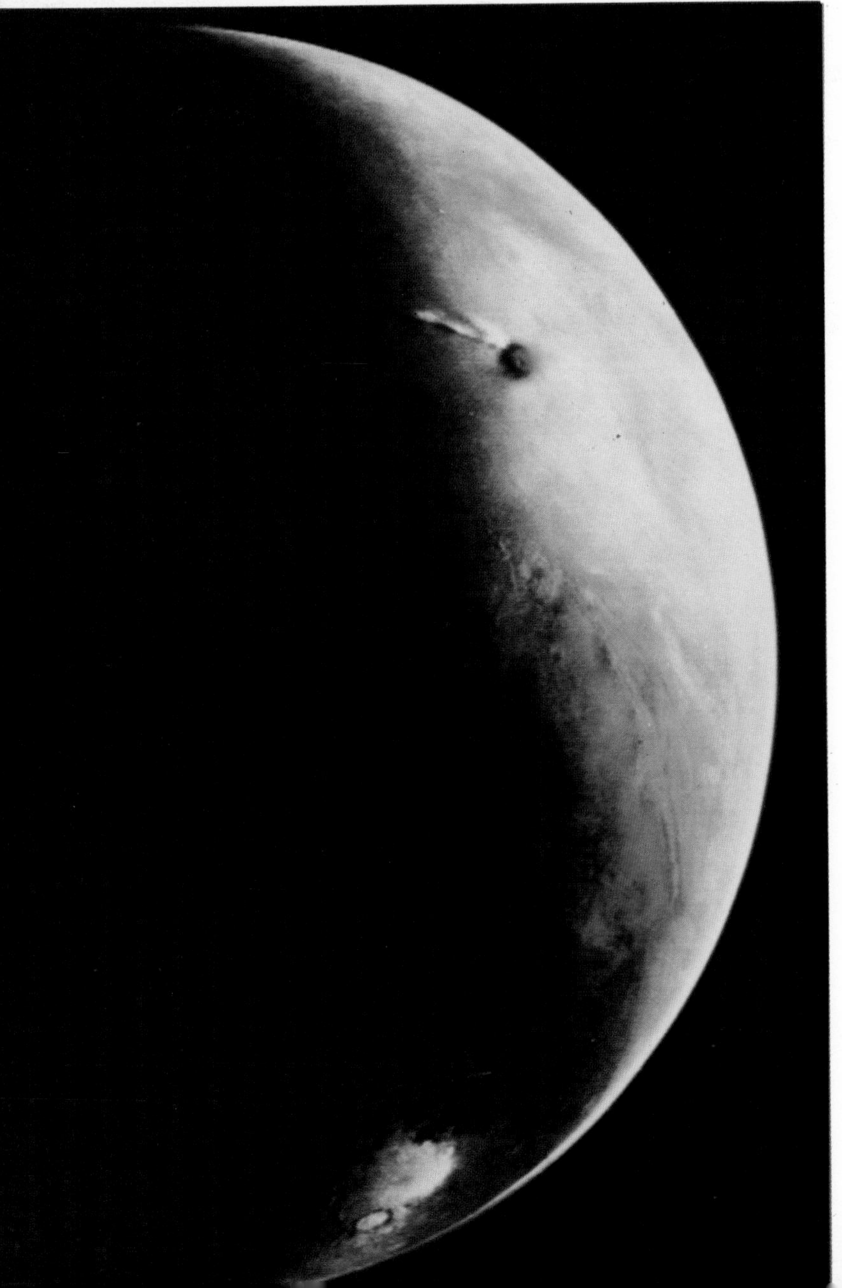

MARS

This global portrait of Mars (TOP LEFT) looking down on the north pole is a mosaic of 1,500 pictures taken by Mariner 9 in 1971 and 1972, the first composite picture of its kind to be made of any body in the solar system.

The Viking program, which followed the Mariner probes, involved the most sophisticated technology yet deployed in space. The near-human spacecraft were piloted by two computers, suspiciously checking on each other's actions, whose order of "seniority" was established by intelligence tests before the flight.

Viking and Mariner provided more information about Mars than we have about any other planet, but they failed in their primary mission – to establish whether there is life there. Some of the elaborate laboratory tests carried out by the Viking landers said yes; others were negative. However, most scientists are now agreed on one thing: if there are no actual organisms in the Martian soil, there are some suspiciously life-like chemicals.

From the stream of data and pictures it is clear that the legendary canals do not exist, but there are signs of water everywhere. In fact, Mars is undergoing an ice age, and most of its liquid assets are locked into the giant polar ice caps. The eroded canyons show that rivers and lakes once flowed on the surface, and one day they will do so again.

This view (BOTTOM LEFT) was transmitted by Viking Orbiter 2 as it flew into the Martian dawn in early August 1976. Although Mars is only half the size of earth, it gives an idea of the stupendous scale of the planet's geography. The plume (top center) is a cloud of ice particles blowing from the summit of one of the giant volcanos, Ascreaus Mons, while the canyon above it is the 1,865 mile (3,000km) long Valles Marineris, named after the spacecraft that discovered it.

MERCURY

The US Mariner spacecraft transmitted this picture (TOP RIGHT) from one of the hottest places in the solar system – the side of Mercury facing the sun.

With no atmosphere to filter the radiation, the temperatures rise to over 800°K. At ground level, down in the Caloris Basin (emerging from the shadow line on the left), the surface is already hot enough to melt lead. Mercury is a heavy metal planet which is 80 percent nickel iron.

Another unexpected phenomenon discovered by Mariner was a magnetic field. Astronomers were surprised at this because Mercury lacks the internal currents of lava, like those inside the earth, which normally produce magnetic fields. It has even been suggested that the metal core might be a huge, weak dipole magnet – although there is no convincing explanation as to how this could have occurred.

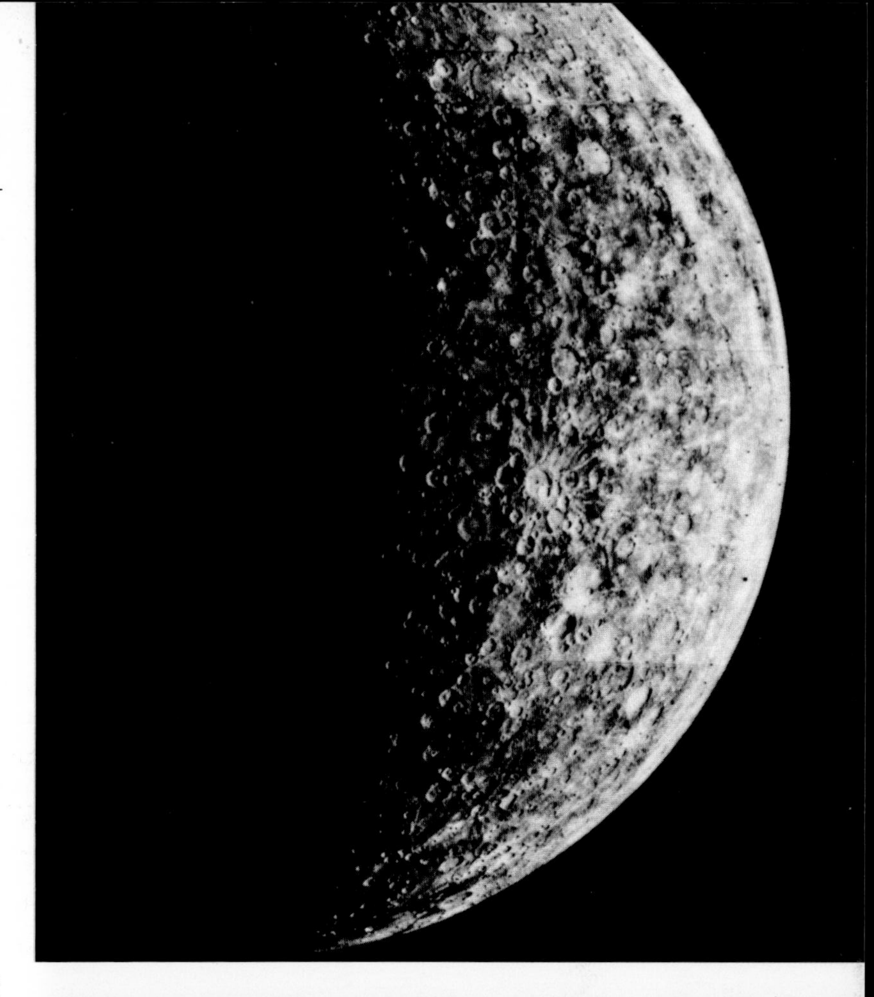

VENUS

Venus (BOTTOM RIGHT) is an enigma. As the US Pioneer spacecraft approached the planet, it grew less and less distinct until from 40,000 miles (65,000km) out it was no more than a hazy outline, like a giant tennis ball unaccountably lobbed into space. The scanners were working perfectly and the pictures were computer-enhanced to reveal the maximum detail. It was the planet itself which was out of focus.

Unlike the structured systems of Jupiter (see pp.56–57), the meteorology of Venus is in a state of featureless chaos. The refractive index of the dirty yellow clouds exactly corresponds to a 75 percent solution of sulphuric acid, and the brew is enriched at lower levels with hydrochloric and hydrofluoric acids to form a scalding, corrosive fog which has destroyed some spacecraft before they even had a chance to land.

Even without the clouds, it would be almost impossible to see the surface from space, because the density of the atmosphere produces optical illusions. One of these is a phenomenon called the Rayleigh Effect. When light passes through a gas, the shorter (blue) wavelengths are more easily scattered than the long (red) ones. The atmosphere of Venus is so dense that surface detail can only be seen in infrared light; and color pictures taken at ground level would show a ruby sun, multiplied by mirages, in a dark magenta sky.

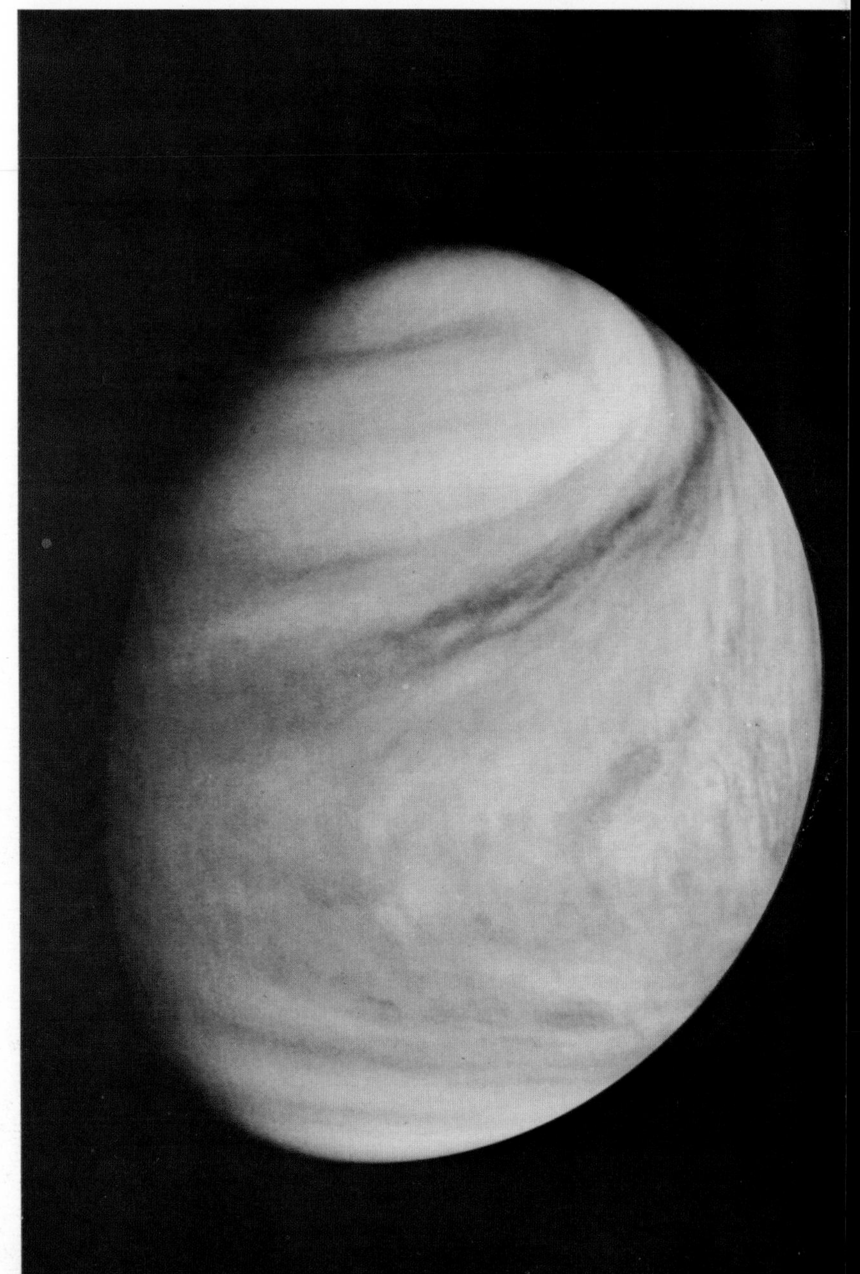

THE OTHER PLANETS

In addition to the nine planets, the solar system contains 37 moons, 3,000 (officially listed) asteroids, countless bodies less than 6 miles (10km) across, myriads of rocks and stones and at least 100 billion comets. Until spacecraft began to gather their rich harvest of images, these were nearly all invisible.

The composite of pictures taken by Voyager spacecraft (OPPOSITE) shows Jupiter, in the background, with four of its fourteen satellites: the small sulphurous boil of Io; Europa, with its unnaturally smooth surface; and the ice planets, Ganymede and Callisto. Uranus and Neptune have seven or eight moons between them, which have yet to be photographed as anything but dots of light, and Saturn another ten, including Titan, which is nearly as big as earth.

The tiny red satellite Amalthea (TOP) races around Jupiter every twelve hours at a height of only 1.5 times the planet's radius. The picture was taken by Voyager 1 in March 1979 and mission controllers considered it something of a fluke. At a distance of a quarter of a million miles, the 80-mile-wide satellite was at the limit of resolution and would only be in the field of view for a moment. To make it even more of a long shot, the cameras were focused on a narrow arc of sky so close to the planet that the surface glare all but obscured the picture. After an impatient half-hour wait for the image to be transmitted and processed, NASA scientists found that Amalthea had been dead on cue. Computer-enhancement did the rest.

Phobos (BOTTOM) is the innermost satellite of Mars and a mere 17.5 miles (28km) across. The computer-enhanced photograph by Mariner 9 shows surprising striations and chains of small craters usually caused by the debris from larger impacts. It is clearly very old, and the irregular edges suggest that it has been knocked off a larger body, which seems to confirm the theory that Phobos and Mars's other moon, Deimos, were once a single planet that split in two as the result of a violent collision.

There are thousands of bodies like Phobos and Amalthea in the solar system. They range in size from 250 miles (400km) down to 6 miles (10km), many with their own set of miniature planets and others orbiting around each other in pairs.

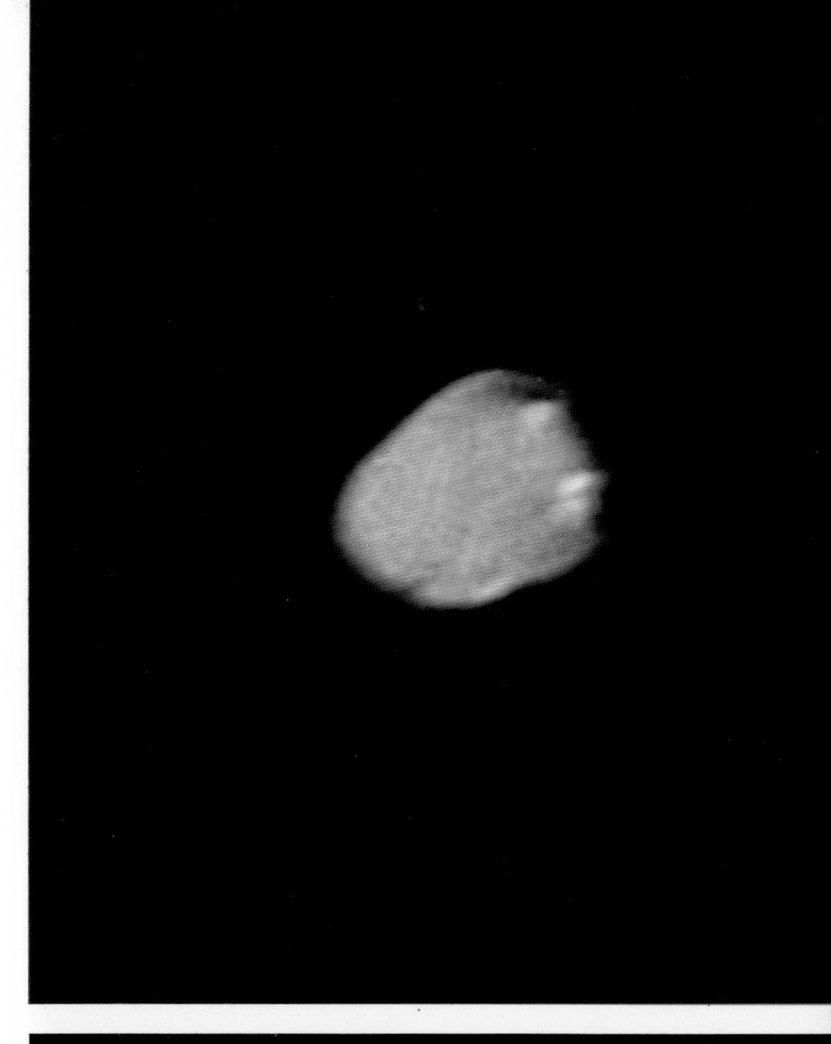

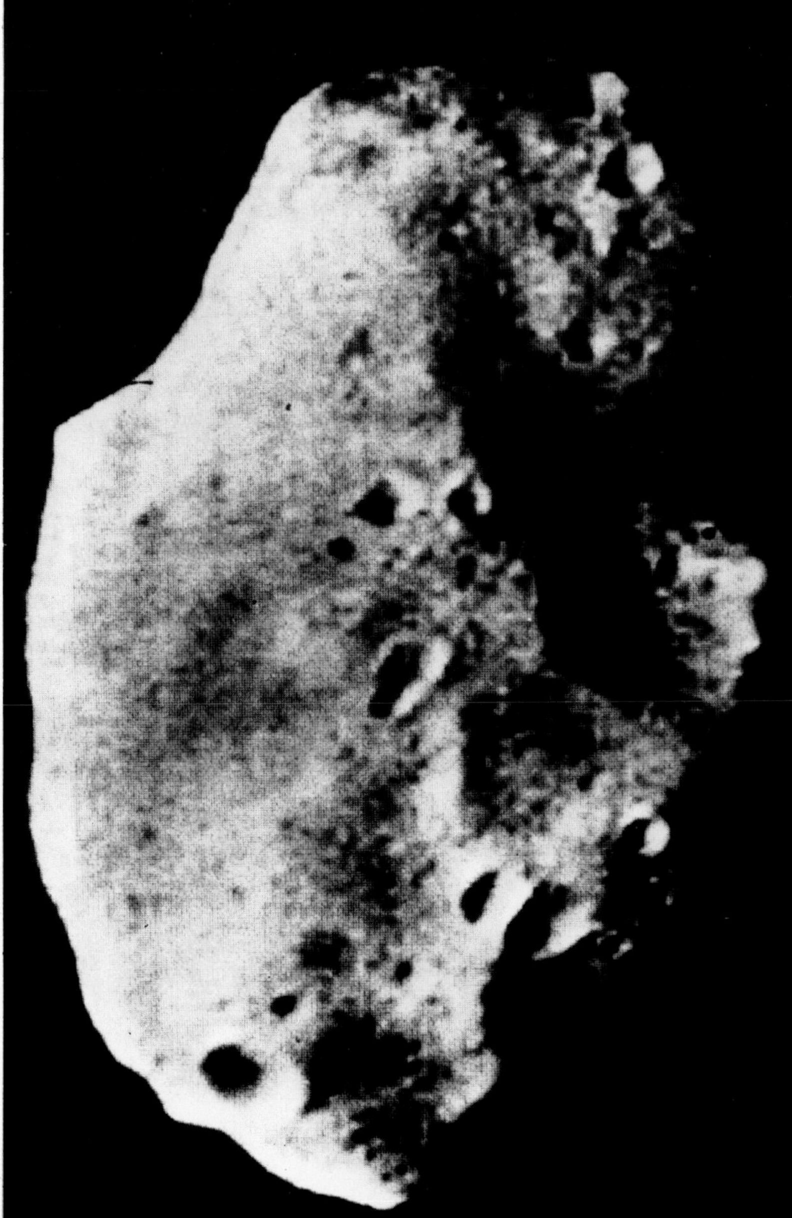

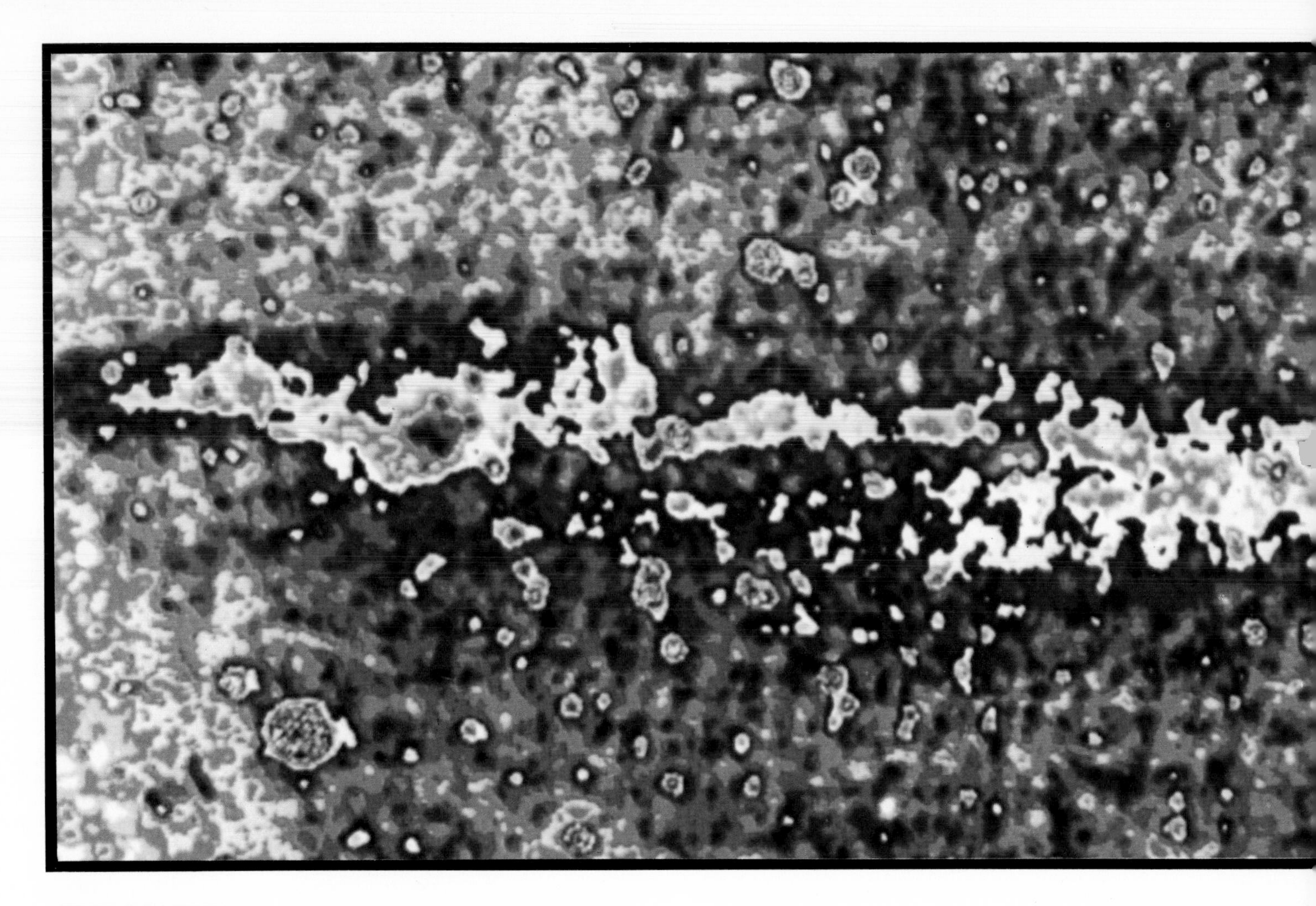

THE COMETS

They called them *cometes*, the long-haired stars, blazing omens from outer space which were liable to crash to earth. But comets are as much a part of the solar system as we are. They are cold, not hot; and far from being solid, they are transparent enough to see stars through.

The nucleus of a comet is composed of an irregular swarm of dust particles and gas, picked up by the solar system as it circles the galaxy. Literally millions of them are trapped in a huge collection of frozen nuclei called Oort's Cloud, in the outer reaches of the solar system.

A few have taken up the familiar elongated orbits associated with comets, which periodically brings them into the inner part of the solar system, to swing around the sun before returning to deep space for hundreds or thousands of years. As they approach the sun, their shape changes, compressing in some directions and extending in others. The frozen gas thaws out, starts to glow and then to fluoresce as the solar wind blows it outward like a rigid torch, while a wake of dust particles is left scintillating in a curved path behind it. The two tails are quite distinct and when the comet moves away from the sun they actually point in different directions, with the particles trailing behind and the slowly fading torch projecting in front.

Only five hundred comets have been recorded, few have short enough orbits to be seen regularly and only one, Comet Kohoutek (1974), has been studied in any detail.

The nuclei of comets are thought to contain complicated chemicals, including fragments of hydrocarbons and other almost-organic molecules. Scientists are urging that a spacecraft should rendezvous with Halley's Comet, when it returns in 1986, to obtain samples of it. Since much of the material originally came from outer space, they may provide the first evidence of extraterrestrial life. It is even remotely possible that the earth itself was "seeded" with life as it passed through the tail of a comet 3 billion years ago.

ABOVE: **The three-million-mile tail of Comet Kohoutek in a computer-enhanced version of a photograph taken from Skylab on Christmas Day, 1973. Kohoutek did not come close enough to the sun to make a really impressive display, as seen from earth, but it was the first time that electronic imaging was used to reveal the internal structure of a comet.**

BOTTOM LEFT TO RIGHT: **Pictures of Brooks' Comet (1893), Finsler's Comet (1937) and Comet West (1976) show the frustrated attempts of astronomers down the years to photograph the "long-haired stars." No matter how good the resolution of the telescopes, the structure of the comet was always obscured by its glowing halo of gas. When photographs of Comet West were computer-enhanced, however, it was a very different story** (OVERLEAF).

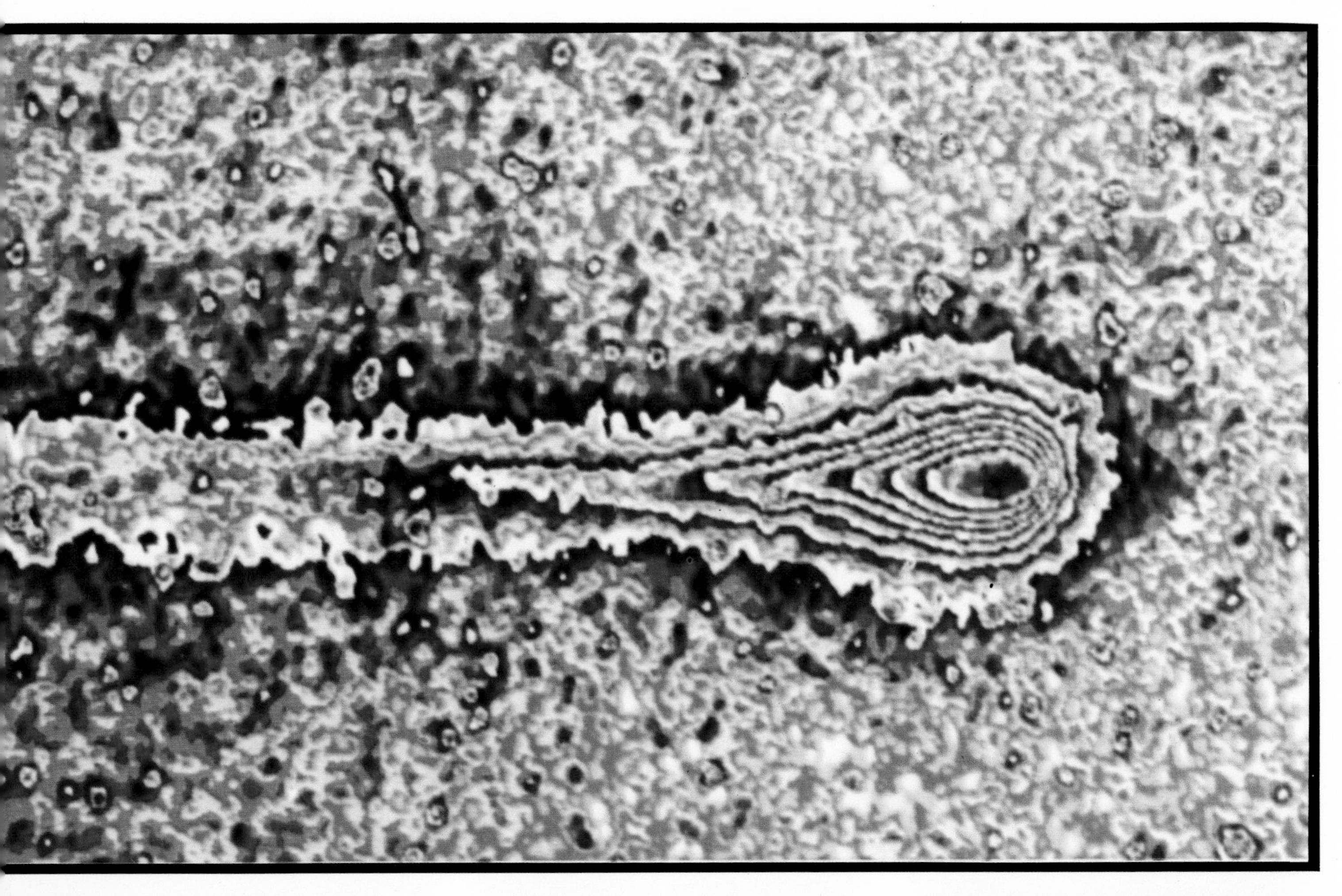

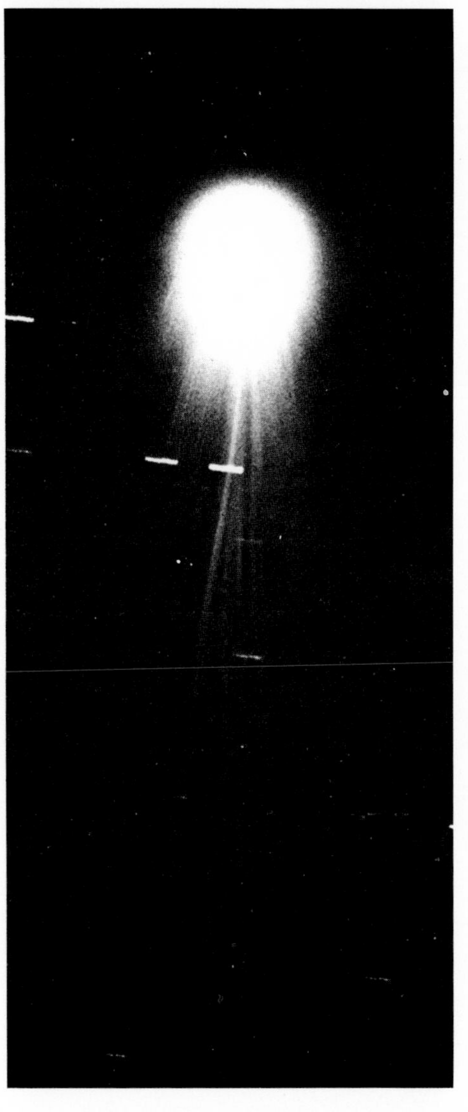

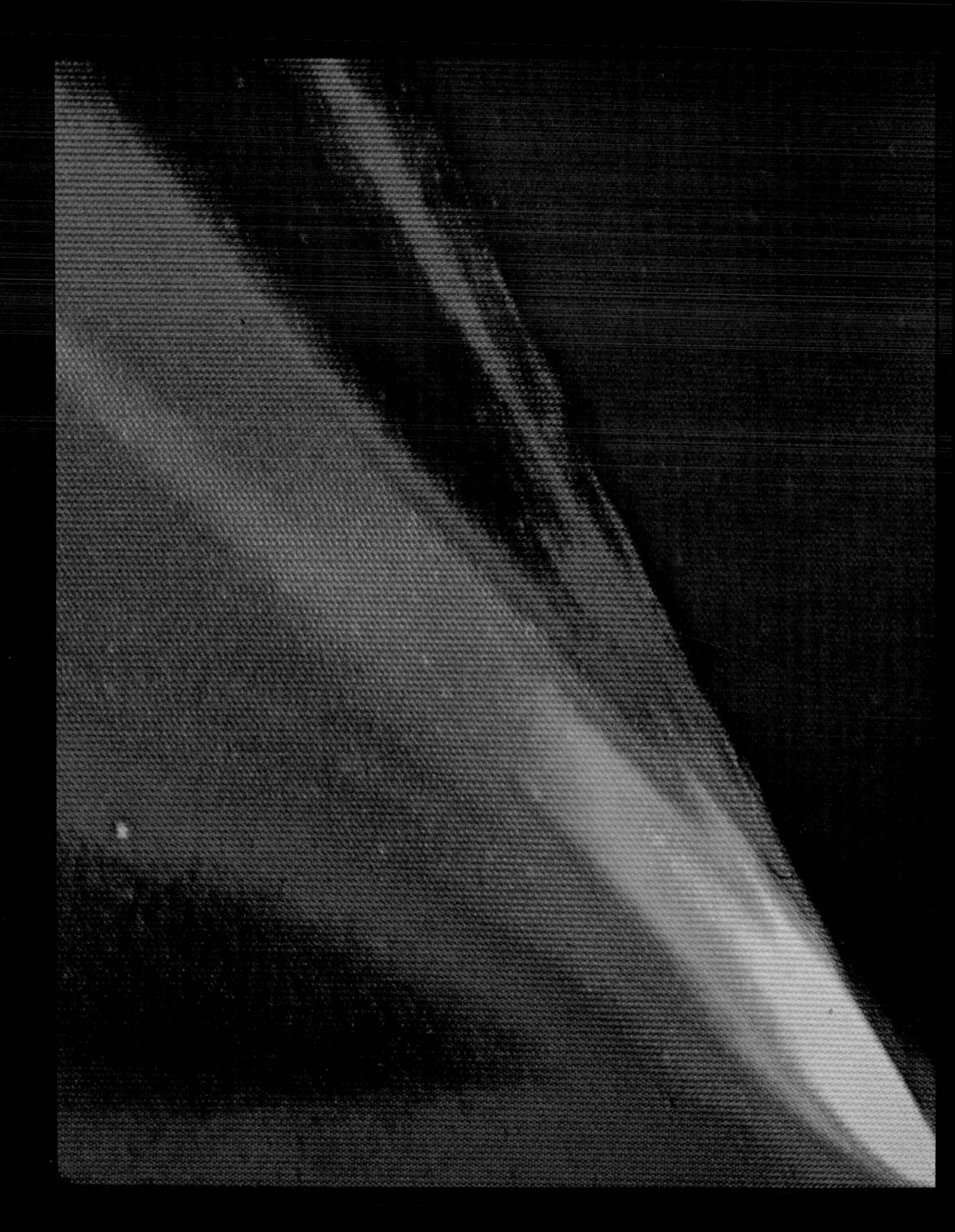

COMET WEST
Comet West was one of the brightest in recent years, with an unusually long tail. Photographs taken by the UK Schmidt Telescope in Australia and computer-enhanced at Kitt Peak showed complex strands and knotty structures trailing from different parts of the nucleus. The comet eventually disintegrated into separate pieces as it rounded the sun.

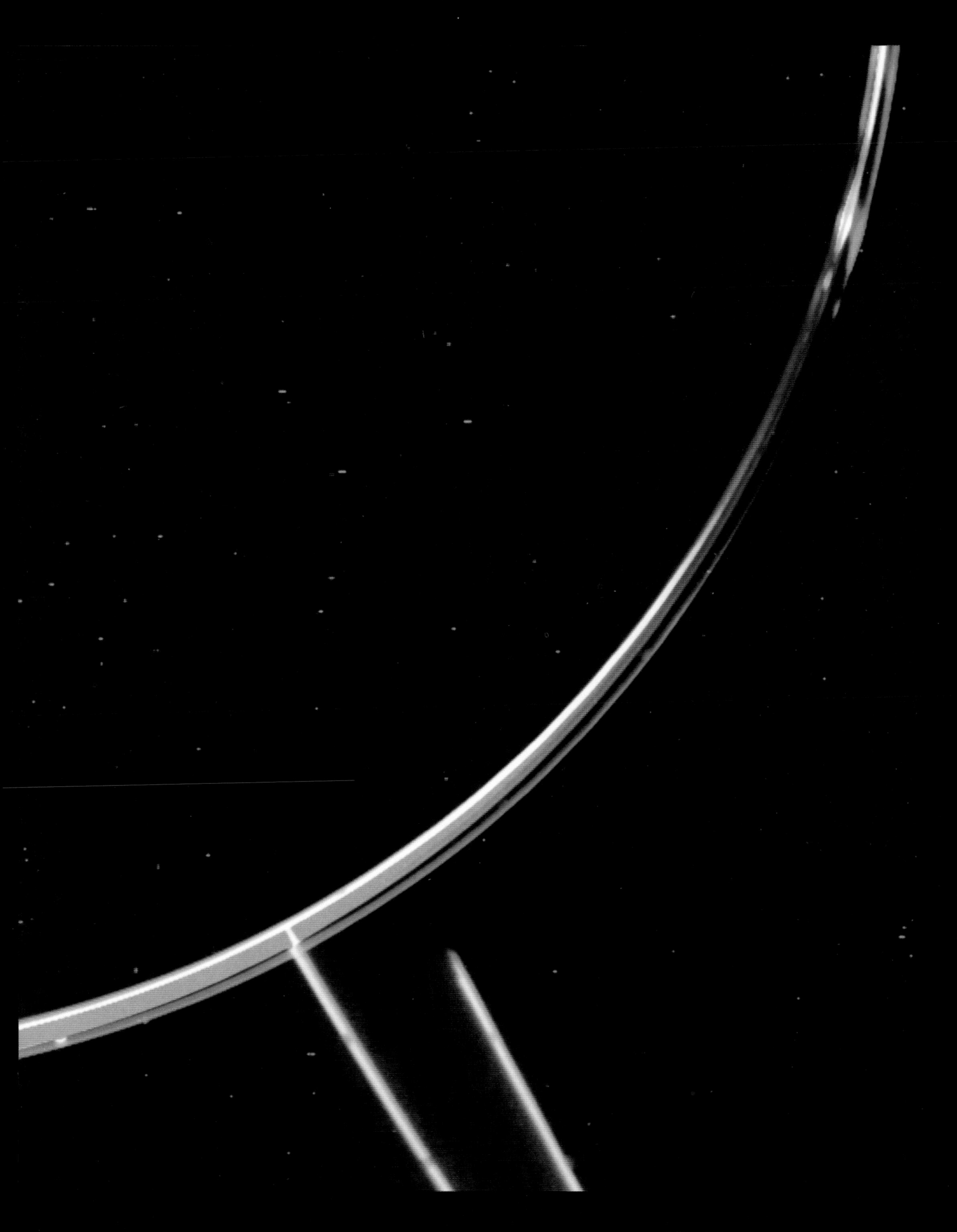

JUPITER'S RING
The back-lit halo of the ring around Jupiter was photographed for the first time by Voyager 2. The picture required such a long exposure that the rim of the planet appears as multiple images. (The specks are artifacts, accidental products of the imaging system.) Unlike the rings of Saturn and Uranus, Jupiter's ring consists of dusty material less than 19 miles (30km) thick, and could well be the debris of a passing comet "captured" by the planet.

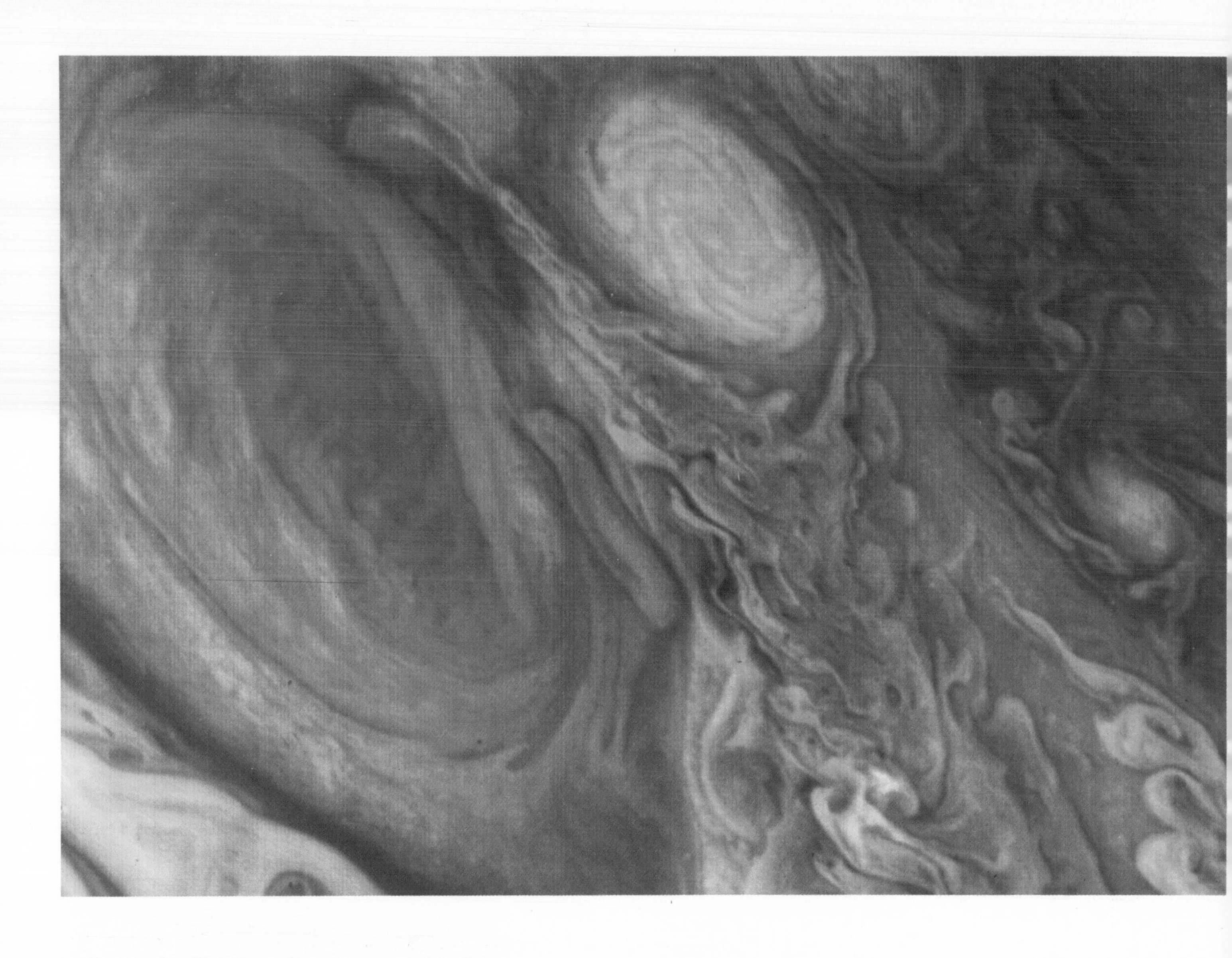

THE THOUSAND-YEAR-OLD STORMS

There was a four-month interval between the visits of the two Voyager spacecraft to Jupiter, and they recorded considerable changes in the weather patterns. But the planet's meteorology is so vast that systems are stable and individual storms can last for centuries.

The atmosphere of Jupiter is thousands of miles thick, compressed to the consistency of water, then chewing gum and finally solid material the deeper one goes. On the surface it is an immense caldron of hydrogen, methane, water, ammonia and other gases in turbulent motion at different speeds, lit by enormous bolts of lightning and auroral arcs 18,750 miles (30,000km) long, which are visible in broad sunlight.

In these pictures the weather is traveling from right to left. The upper levels are moving faster than the lower ones, and the narrow bands of white clouds are compressed jet streams circling the equator at up to 330 feet (100m) a second. The famous "red spot" is trapped in this fast easterly flow, partly blocking it as it spins anticlockwise.

The Voyager 1 close-up of the "red spot" (ABOVE) has been color-enhanced (OPPOSITE) by emphasizing the red and blue components at the expense of the green, so that more details of the internal structure are visible. The distance from top to bottom of the picture is about 15,000 miles (24,000km) and the smallest features visible are about 20 miles (30km) across.

The "red spot" is an anticyclonic storm (similar to those on earth, though much larger) which was first recorded about three hundred years ago. However, recent calculations show that it could be as much as a thousand times older. In early Victorian times the spot was no more than a pale pink blob, but in 1878 it began to swell and darken. It reached its largest recorded size, equivalent to the whole area of earth, about ten years later. It underwent another decline in the 1960s, but has since recovered and now measures about 13,000 miles (21,000km) from side to side.

Although it remains at the same latitude, the spot is not related to anything on the surface and constantly moves around. The dark red color is caused by phosphene gas (which smells of garlic) welling up from below. At the surface ultraviolet radiation from the sun breaks it up into various phosphorous compounds. It looks like a deep vortex, but in fact the material in the center is several hundred miles higher, and colder, than the clouds around it. The colors are warm, but the cloud tops themselves are a chilly −110°C.

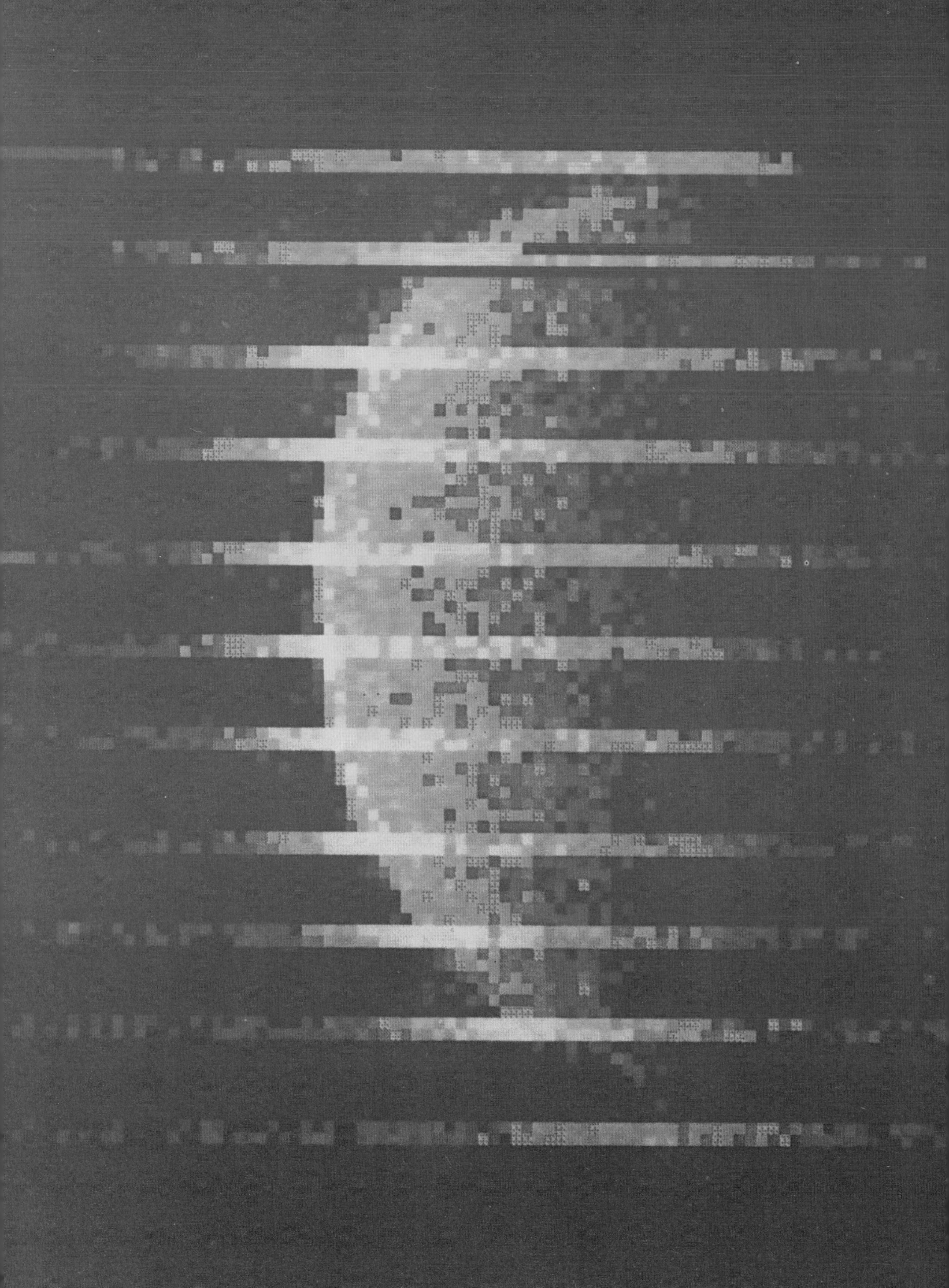

HIGH WIND ON VENUS

The atmosphere of Venus is simpler and less structured than Jupiter's, but no less violent.

The double image of the planet (OPPOSITE), transmitted by the Pioneer orbiter, shows how electronics can combine different types of information. The inner crescent is a picture of Venus made in ultraviolet light. The strips are brief interruptions to the signals to measure the light emitted by atomic hydrogen, which shows that thin clouds of gas extend for thousands of miles into space. Between these limits, at a height between 20 and 50 miles (30 and 80 km), is the raging maelstrom of the Venusian atmosphere.

The violent meteorology is the result of a "greenhouse effect," an ecological disaster which may threaten earth if we allow the carbon dioxide in our atmosphere to reach too high a level. At a certain point the gas acts as a filter, letting in the sunshine but preventing the heat from radiating away. The result on Venus is a continuous raging storm with winds up to 330 feet (100m) a second and clouds moving at speeds up to 60 times faster than the surface is rotating.

In contrast to this activity, the surface itself is an eerie place with little wind to disturb a dark, hot fog, one-tenth as dense as water. Only 1 percent of the sun's light reaches the surface, but the sulphuric acid vapor is at a steady 650°C with a pressure 100 times that on earth.

The weather is the same all over the planet. There is less than 2 percent difference in temperature between the equator and the poles, compared with a 40 percent difference on Mars and about 15 percent on earth.

When one of the USSR Venera spacecraft finally made a soft landing and survived for long enough to send back pictures, they showed a landscape without any signs of erosion. The cratering is minimal, because most meteors burn up in the atmosphere. Unlike Mars there are few dust storms, and Venus has never had enough water to collect on the surface. There has been nothing to interfere with the natural geological forces at work piling up mountains and growing continents in the hostile darkness.

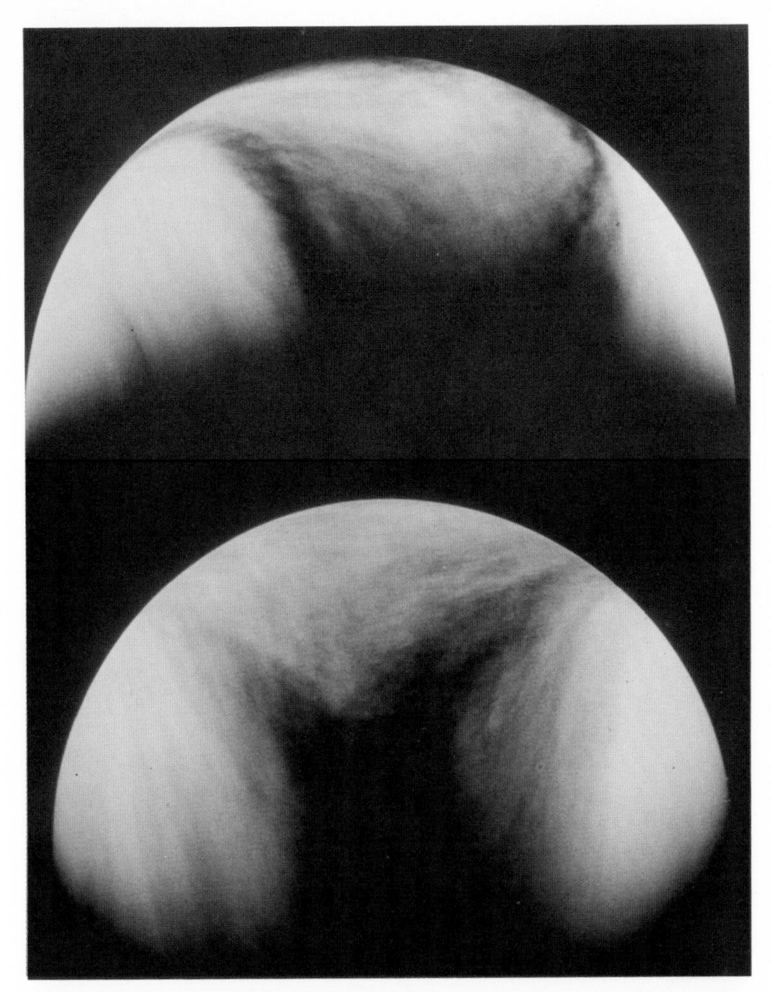

TOP RIGHT: **The speed and violence of the Venusian atmosphere is illustrated by successive pictures from Pioneer. They show that cloud formations in a massive Y-front over the pole (top) had reached the equator by the time it was photographed sixteen days later (bottom).**

BOTTOM RIGHT: **Layers of dust and cloud on the horizon outline the atmosphere of Mars as the Viking Orbiter swings over the frost-covered crater toward the south pole. The air is thin – a mainly carbon-dioxide atmosphere equivalent to that 22 miles (35km) above earth – and the surface is extremely windy, with constant dust storms.**

THE CRATER MYSTERY

Apparently random patterns of cratering have turned out to be an archeological record of the earliest events in the solar system.

When the Mariner spacecraft recorded the first close-up pictures of Mercury (BELOW), they showed the same bleached desolation as the Apollo photographs of the moon (OPPOSITE) – craters and still more craters, as if every landscape in space was the same repetitive, pock-marked desert. But that was before the pictures came in from Mars and Jupiter, and the full story of the craters themselves began to unfold.

Meteorite craters are not just dents in the surface. Where there is no atmosphere to slow them down, they strike at a speed of tens of thousands of miles an hour with such force that they bury themselves deep underground. In some cases, as on Callisto, they disappeared right into the core of the planet. But they usually come to rest a mile or so below the surface, vaporizing the rock around them and causing an underground explosion. In low-gravity conditions the debris is hurled over a considerable distance, producing the typical pattern of an outer ring some way from the point of impact and a central peak. On bodies like the moon, which have a molten core, lava then usually seeps out to cover the immediate area of the impact.

Evidence from the moon, Mercury and Mars shows that everything in the solar system, including earth, went through the same meteor bombardment at a specific point in history. Until recently it was thought that the meteors fell in a great shower about 4 billion years ago and that these impacts gradually reduced until they are now very rare events. The earth craters have long since eroded away, except for a few in Canada and Russia, and in the bed of the North Sea, but the perfectly preserved crater record on other planets has revealed what actually happened.

It seems that there was an even larger meteor shower half a billion years earlier, immediately after the earth, moon and most of the solar system were formed – presumably as a tidying-up process for the material left over.

Why this wave died away completely before the second occurred is a complete mystery. Where were the meteors all this time? Were they stored in orbit, like the present Asteroid Belt, and then suddenly released for some reason? Or did one of the original planets disintegrate? Maybe the solar system ran into a shower of them wandering in space – an event that would be an unthinkable disaster if it happened again.

BELOW: **The surface of Mercury looks similar to the moon but, as well as the fact that it receives ten times as much solar radiation, there are differences. For instance, Mercury's cooling and shrinking metal core has wrinkled the outer crust to produce scalloped cliffs, called lobate scarps, which do not appear on the moon; and the craters are smaller because a stronger gravitational force reduces the spread of debris.**

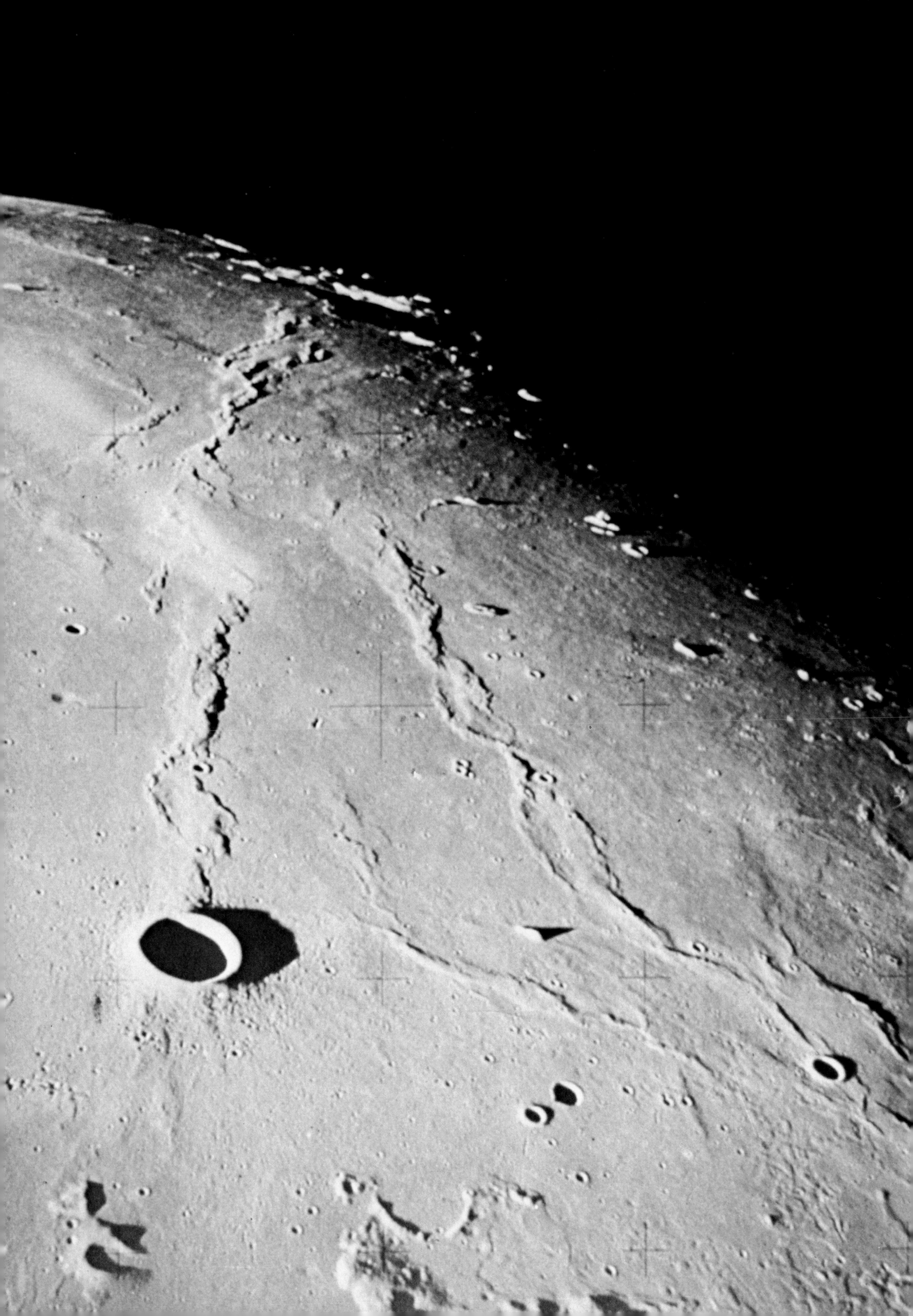

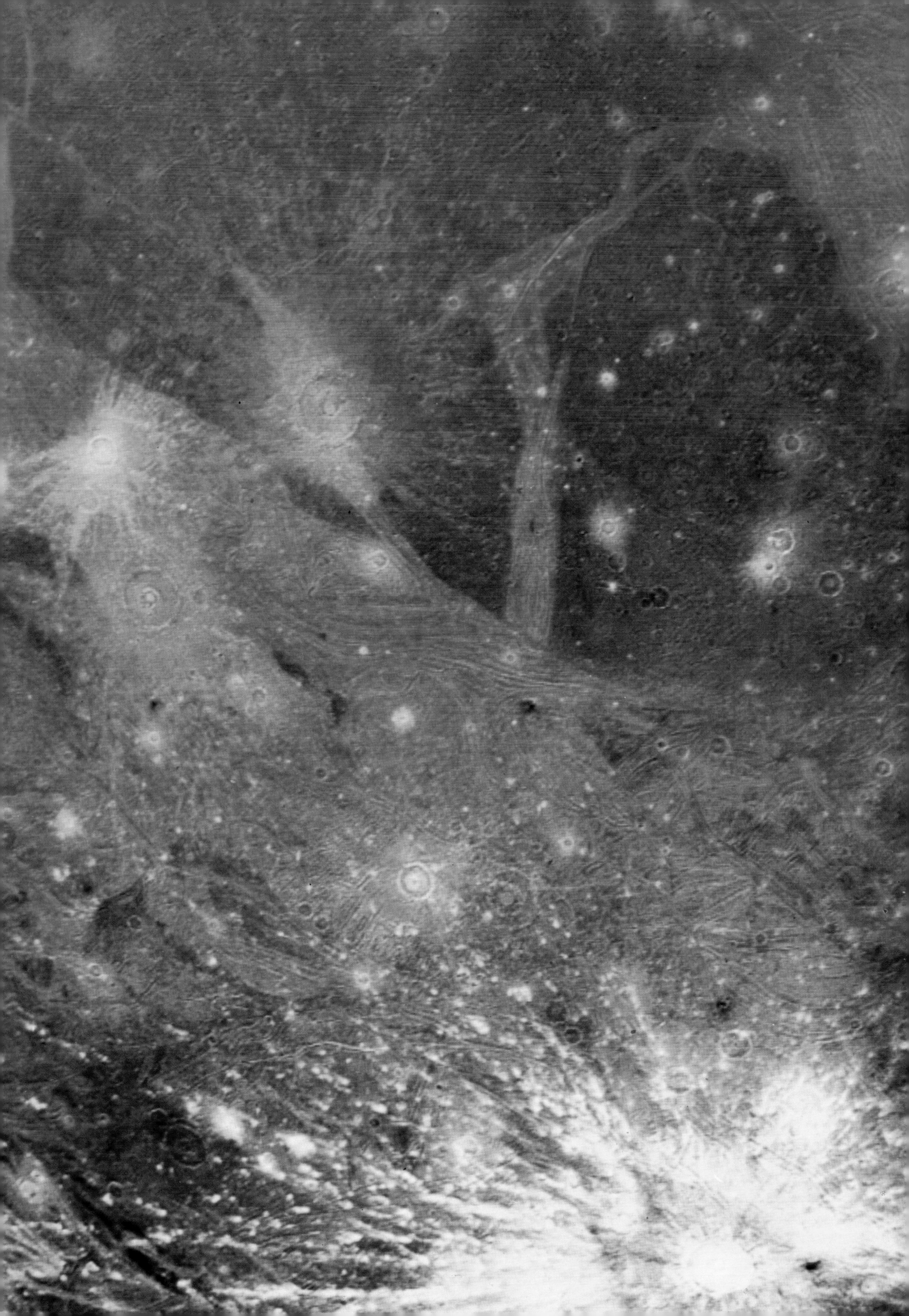

THE ICE PLANETS

Meteorite geysers and melt-down on the surface of Jupiter's frozen moons.

Ganymede (OPPOSITE) is a lump of frozen mud, methane and ammonia, and the effect of a meteor crashing into it is very different from an impact on the moon. The heat is rapidly absorbed as the ice melts, and the explosion is more like a geyser of hot liquids, gas and icy slush, spraying the surface around and freezing as it falls.

The picture, acquired by Voyager 2 in 1979, shows the debris of "ray" craters, like white stars against the darker, older ice.

Planets tend to darken with age, the result of eons of exposure to space dust and micro-meteorites, and the mottled appearance of the moon is caused by a similar patina on its older surfaces. All objects in space gather dust like this, and it is estimated that about ten tons of it settles, unnoticed, on the earth's surface every day. We may be walking on the fragments of other worlds.

Callisto (TOP) is older than Ganymede and has picked up a spectacular record of craters. One of the first views of Callisto from Voyager 1 (BOTTOM) revealed the most unusual results of a meteor impact ever recorded in the solar system. The impact must have been a big one because the heat it generated melted half the planet. The rings in the picture are the remains of tidal waves of mud and ice, 2-3,000 yards high, which radiated outward for 1,660 miles (2,500km) gradually reducing to a groundswell that eventually refroze into these giant ripples.

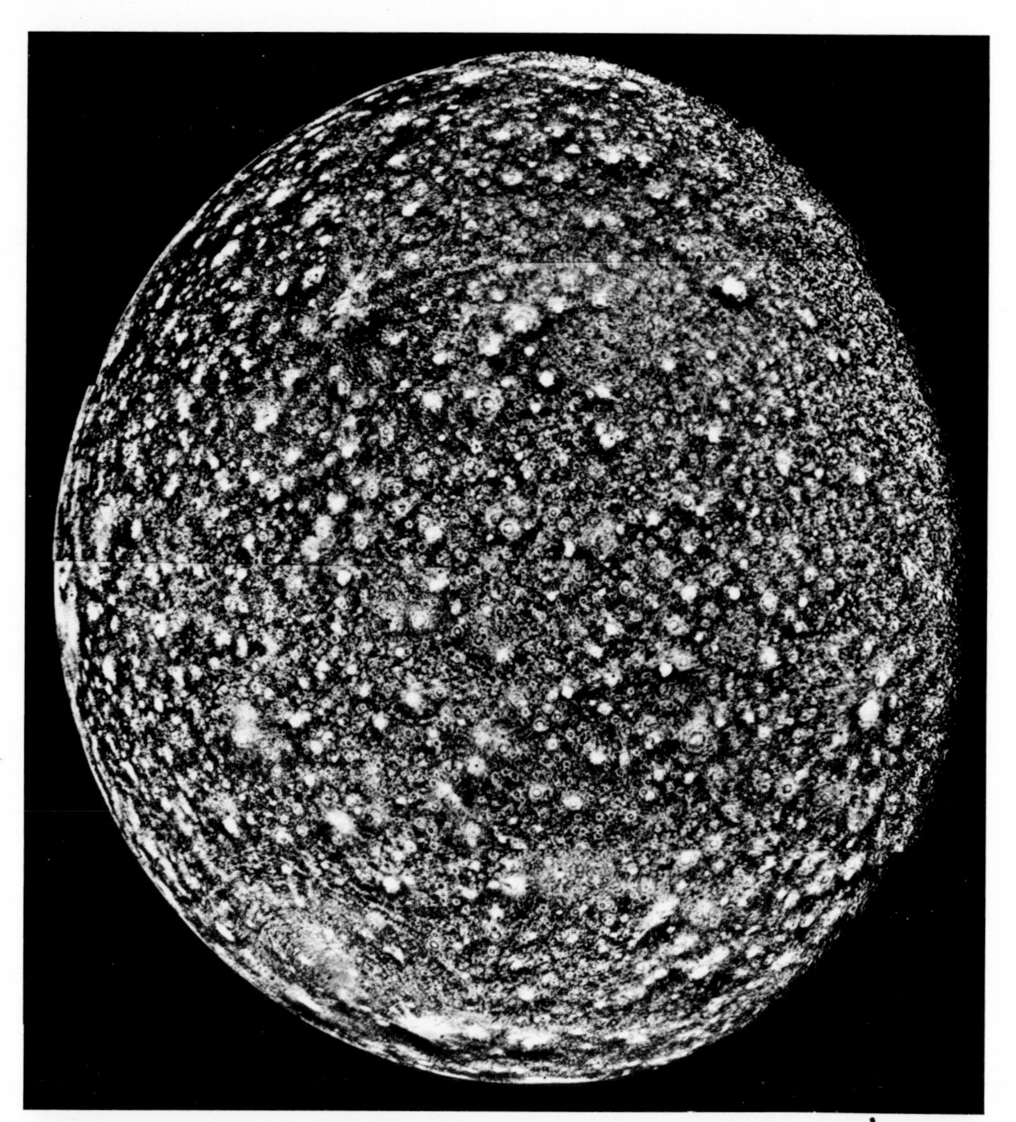

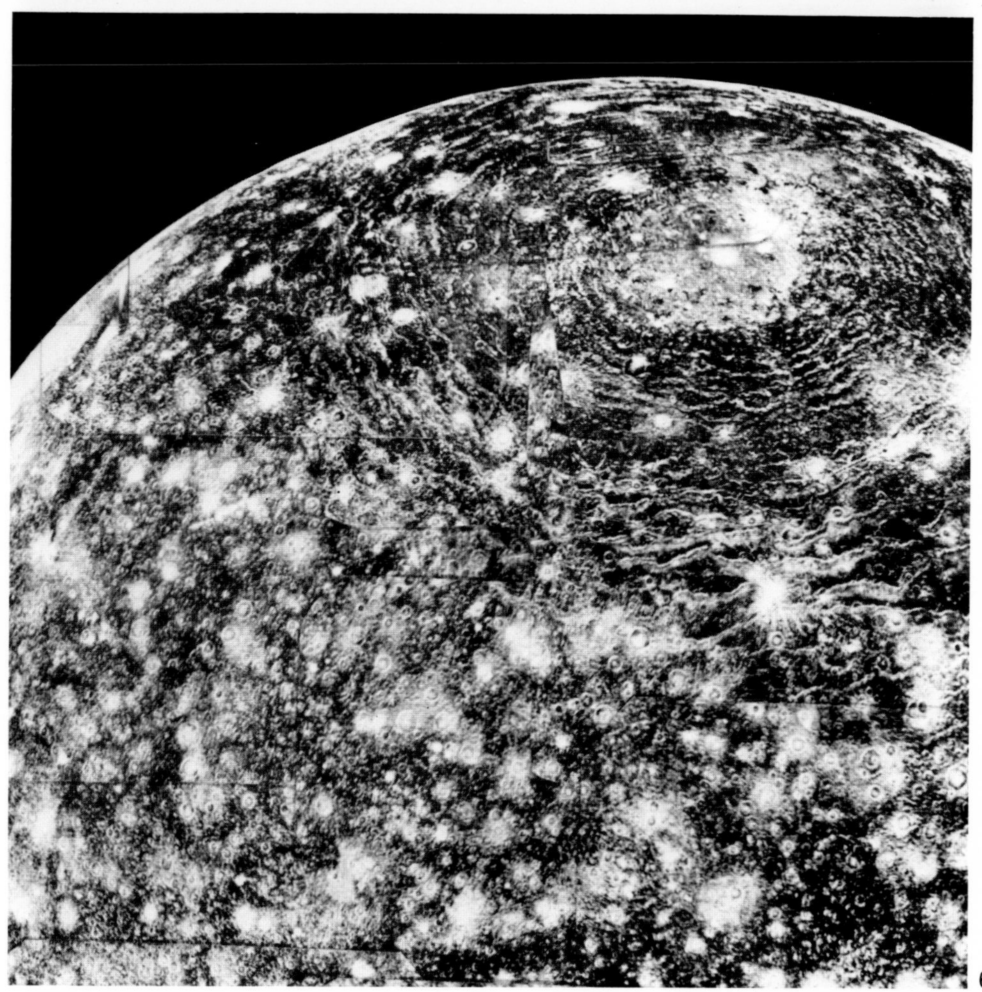

THE VOLCANOES

In contrast to the colder, cratered worlds, the landscape of Jupiter's second moon is like a medieval vision of Hell.

The multicolored lava flows of Io (OPPOSITE) have obliterated all signs of cratering. The Voyager spacecraft recorded no less than eight volcanoes in simultaneous eruption, including one astonishing explosion (ABOVE, RIGHT) three times the size of Krakatoa, all on a body slightly smaller than the moon.

The reason for all this activity is one of the oddest energy systems in space. As it circles the planet, Io cuts through Jupiter's powerful magnetic fields, which are 20,000 times stronger than the earth's. The induction which is set up causes Io to act as an electrical generator. In effect, it becomes an orbiting power station with an estimated current of 5 million amps flowing down a "tube" of magnetic flux to the surface. At the same time, about a trillion watts of heat have to be dissipated in and around the satellite. When this is combined with the tidal pull of Europa and Ganymede, the system ensures that Io is kept permanently on the boil, leaving a wake of sulphurous gases in space behind it.

Most planets go through a volcanic stage, and the material that is ejected makes an important contribution to the atmosphere. A number of elements in the earth's atmosphere, including water vapor, were acquired in this way.

When the volcanoes of Mars and Venus were erupting, they were even more spectacular than those on Io. This is a view (BELOW RIGHT) looking vertically down into the giant basin of the Olympus Mons on Mars, an extinct volcano three times higher than Mount Everest and nearly 400 miles (640km) across.

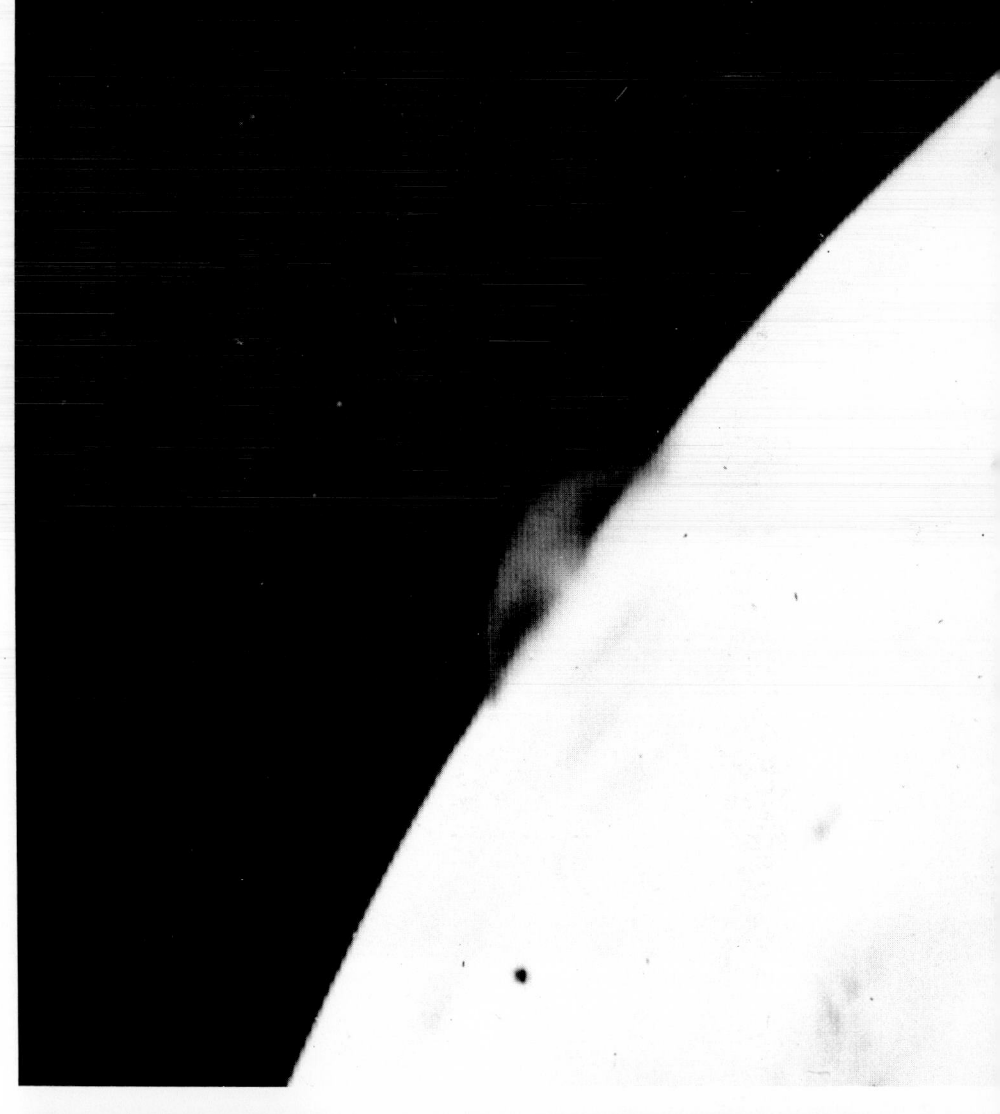

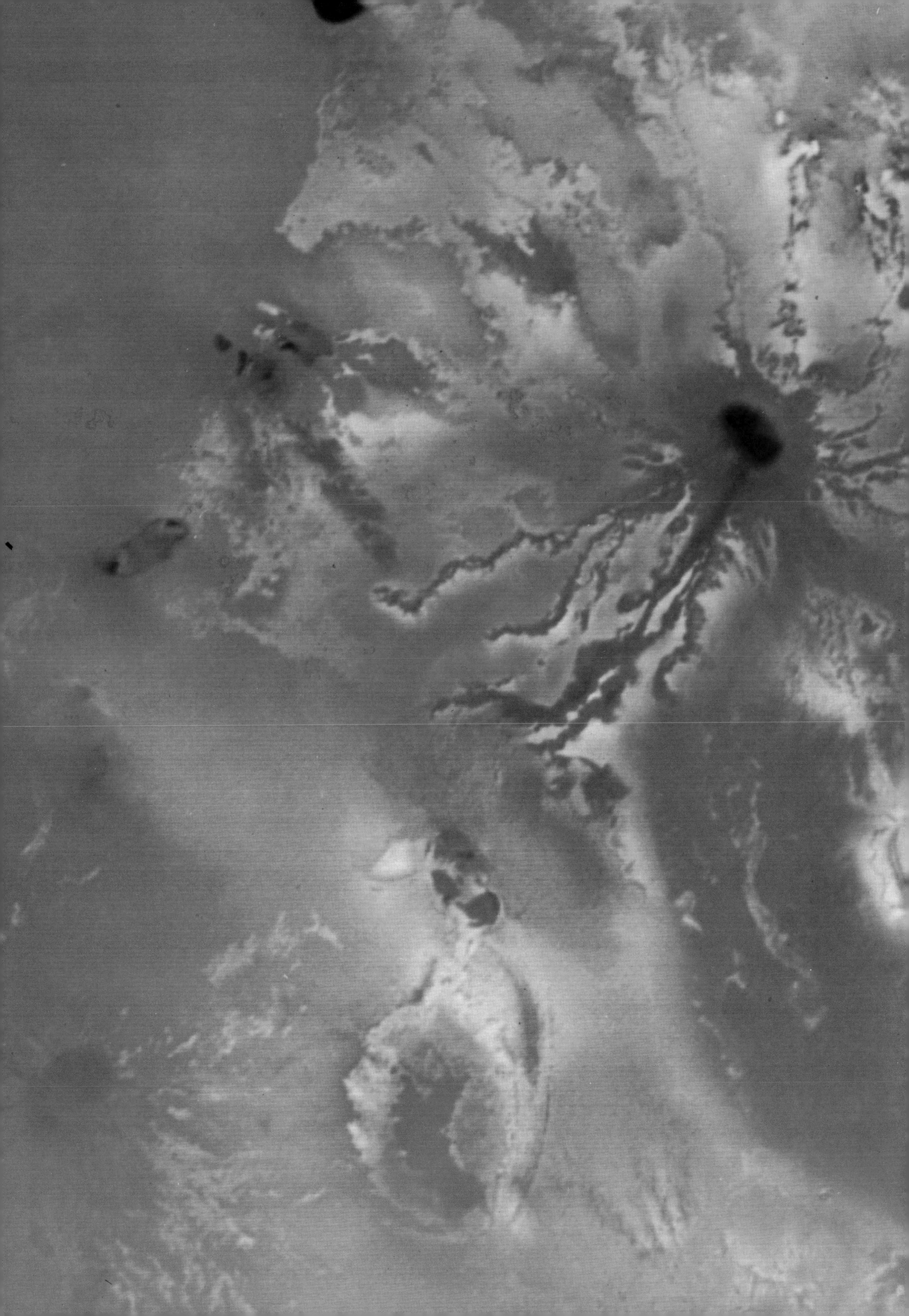

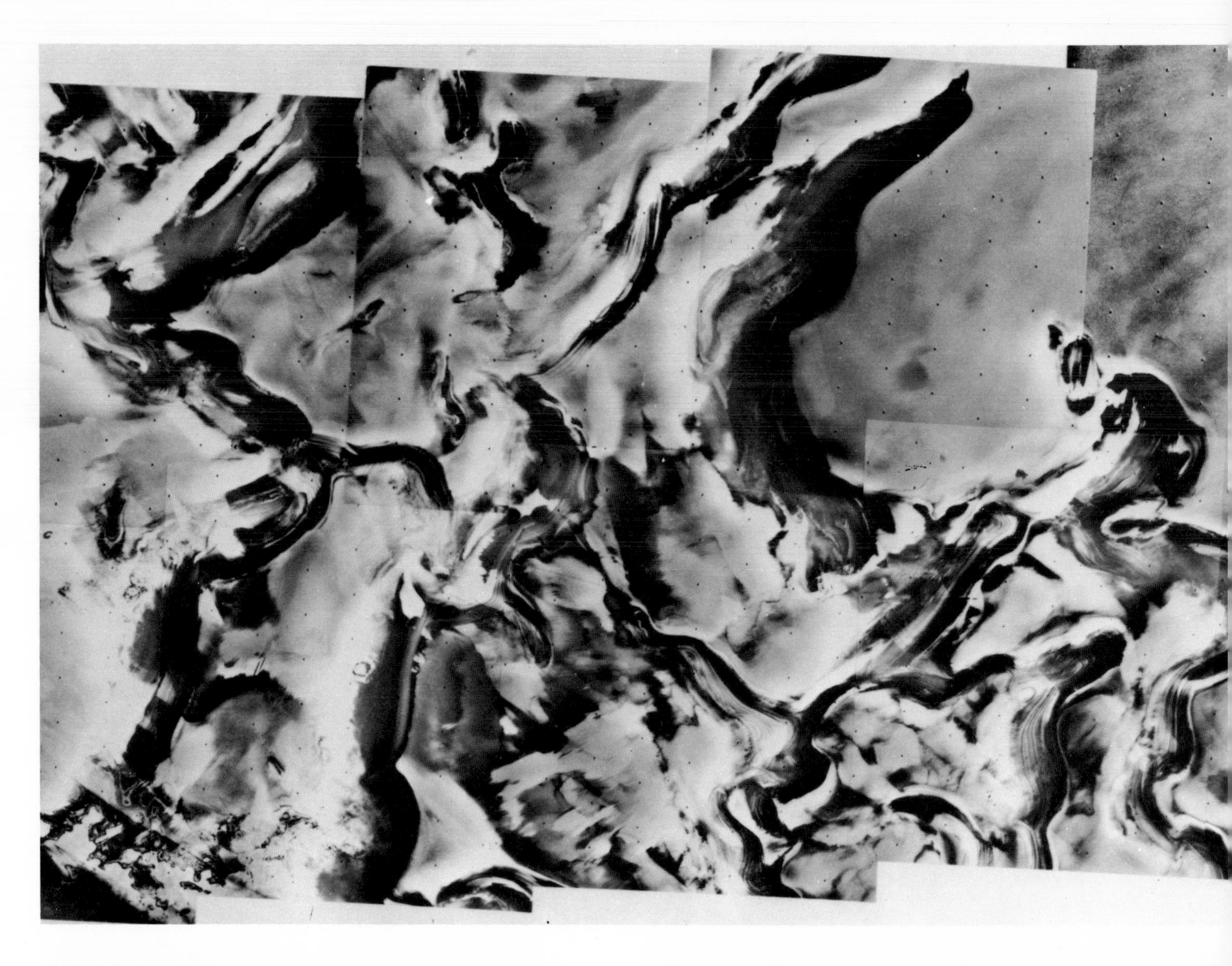

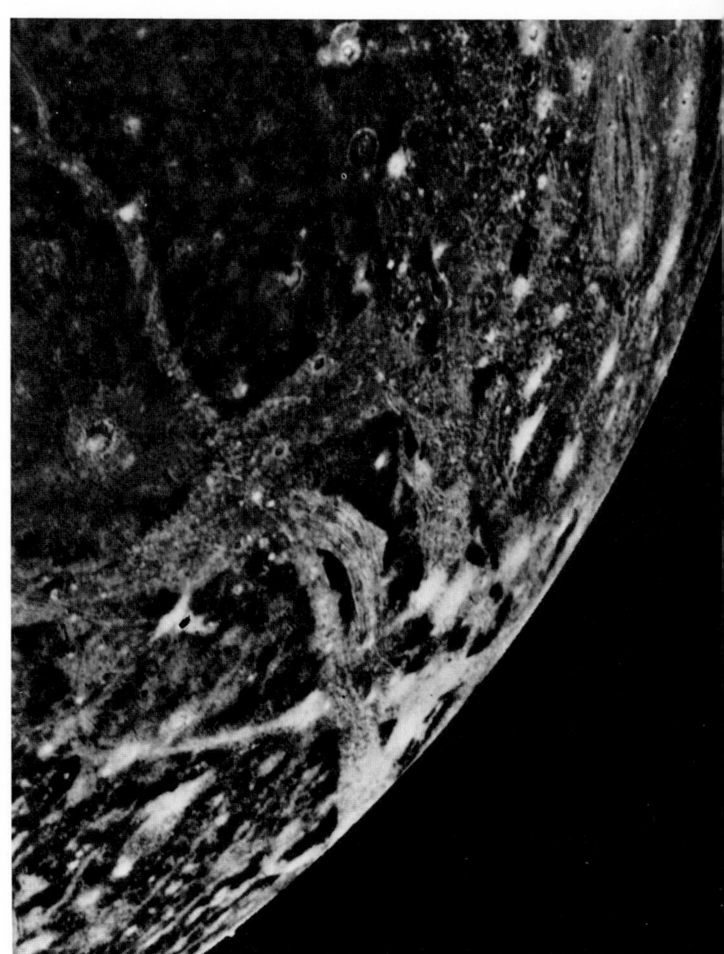

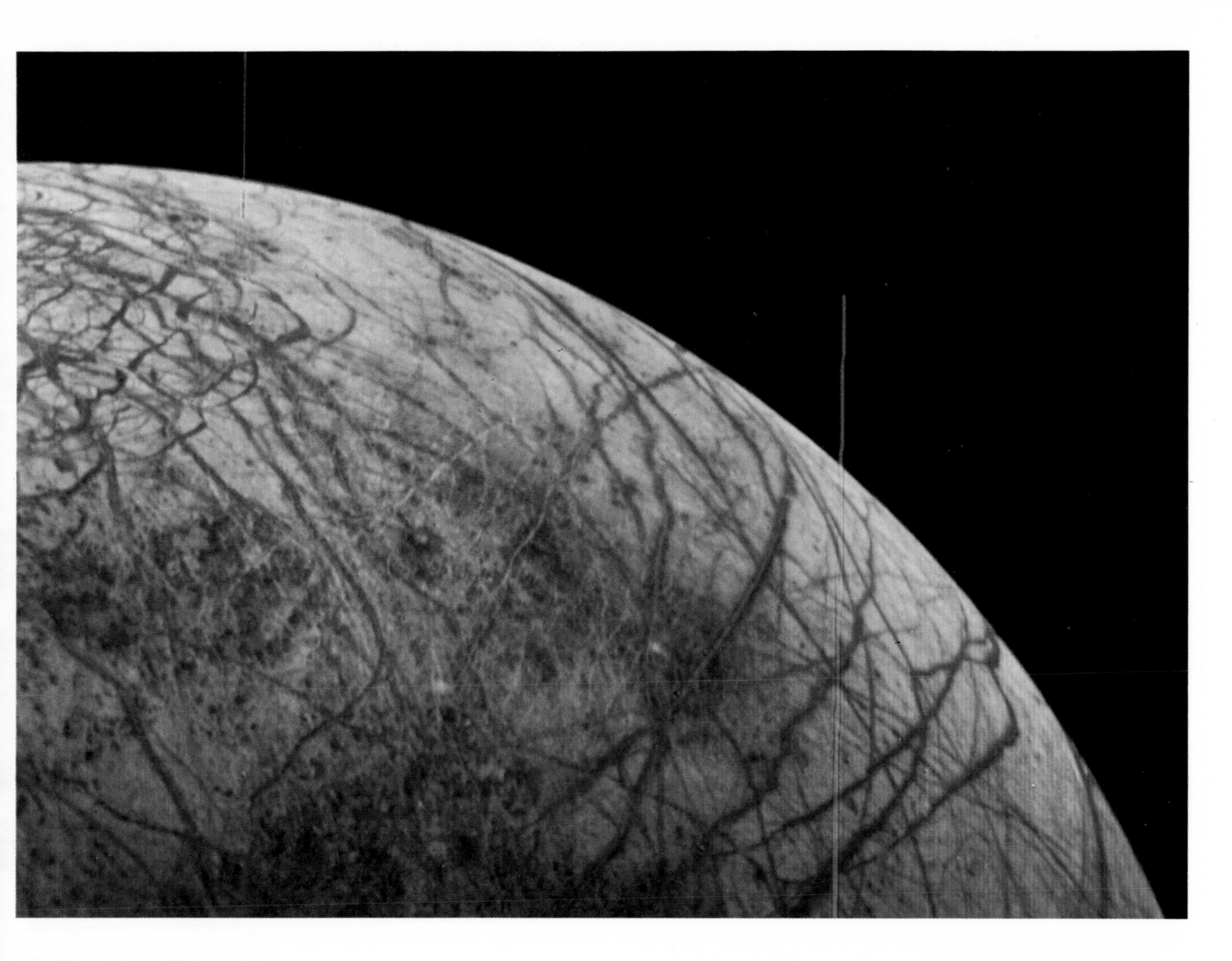

ALIEN LANDSCAPES

Craters and volcanoes are reasonably familiar, but there is literally nothing on earth to compare with the icefields of Mars and Ganymede or the eerie smoothness of Europa.

The swirling chrome-like patterns (OPPOSITE, TOP) are an icefield near the north pole of Mars. Until they showed up on the high-resolution pictures of Viking Orbiter 2, no one had seen anything like them. They are probably the result of dust particles settling on ice that has thawed and refrozen a number of times. Estimates of the amount of water locked up in the polar caps is being adjusted upward. It was once thought that the ice caps were no more than a layer of frozen carbon dioxide, but underneath them there may be sheets of water-ice several miles thick.

If glaciers can be compared to rivers, Ganymede (OPPOSITE, BOTTOM) is wrapped in the tortuous currents of an ice ocean.

These pictures are among the most surprising sent back by the Voyager mission. Like Callisto, Ganymede is at least 50 per cent ice, but whereas it is possible to account for the impact ripples on Callisto a new theory of ice dynamics will be needed to explain these extraordinary stress patterns.

The "shadows" which give an illusion of depth are actually intersecting bands of darker material lying on the surface.

The landscape of Europa (ABOVE), the third moon out from Jupiter, has no mountains or glaciers – or indeed any other surface features. In fact, it is so unnaturally smooth that even at high resolution it is impossible to detect the slightest bump or irregularity.

The reason for this is that Europa is the youngest member of the solar system. It was only formed between 10 and 100 million years ago and the surface has not yet hardened. In order to produce mountains a planet must have a solid crust. The surface of Europa, however, appears to have the consistency of slush, with lava welling up along broad bands which criss-cross the surface.

Europa was one of the first of the minor planets to be discovered by Galileo, in 1610, yet it has taken us till now to realize this single most important fact about it. The same applies, of course, to all these images. It is difficult to grasp the extent to which our ideas have changed, and how banal the guesswork now seems compared to the reality.

THE LANDSCAPES OF MARS

"Look at this," said Carl Sagan to his fellow scientists at NASA, when the first pictures arrived from the surface of Mars (ABOVE). "Bring your full concentration to bear on it. Imagine you are there. What do you see?"

It is 7.30 AM, local time, on August 3, 1976. The sun has only been up a couple of hours and the early-morning shadows still lie across the sand dunes of Mars as the Viking lander starts work. Although it is 220 million miles away, and no human foot has ever set foot on it, this is one of the best-known stretches of ground in the universe. Every square inch has been photographed and rephotographed, distances measured, the soil analyzed and the weather recorded hour by hour. It is so familiar to NASA scientists that even individual rocks have been given names.

In this photograph the camera was looking east across the rolling plain called the Chryse Planitia, which is criss-crossed by the dried beds of rivers which once flowed across this landscape. Unfortunately the lander settled down between them, at a spot where the dunes have remained undisturbed for so long they have partly solidified. The wind that created them also exposed some of the bedrock (middle distance, on the right) and a type of hard soil, like baked mud, called duricrust (bottom left).

To give a sense of scale, the large rock on the left is about 10 feet (3m) long and roughly 25 feet (8m) from the camera. The object in the middle of the picture is the meteorology boom supporting Viking's miniature weather station. The weather at the moment is mild, with no sign of sand storms, temperatures well below freezing and intense ultraviolet radiation. The geography of Mars is gargantuan and although many of the features resemble those on earth, the absence of erosion in the thin Martian atmosphere has allowed them to become hugely magnified.

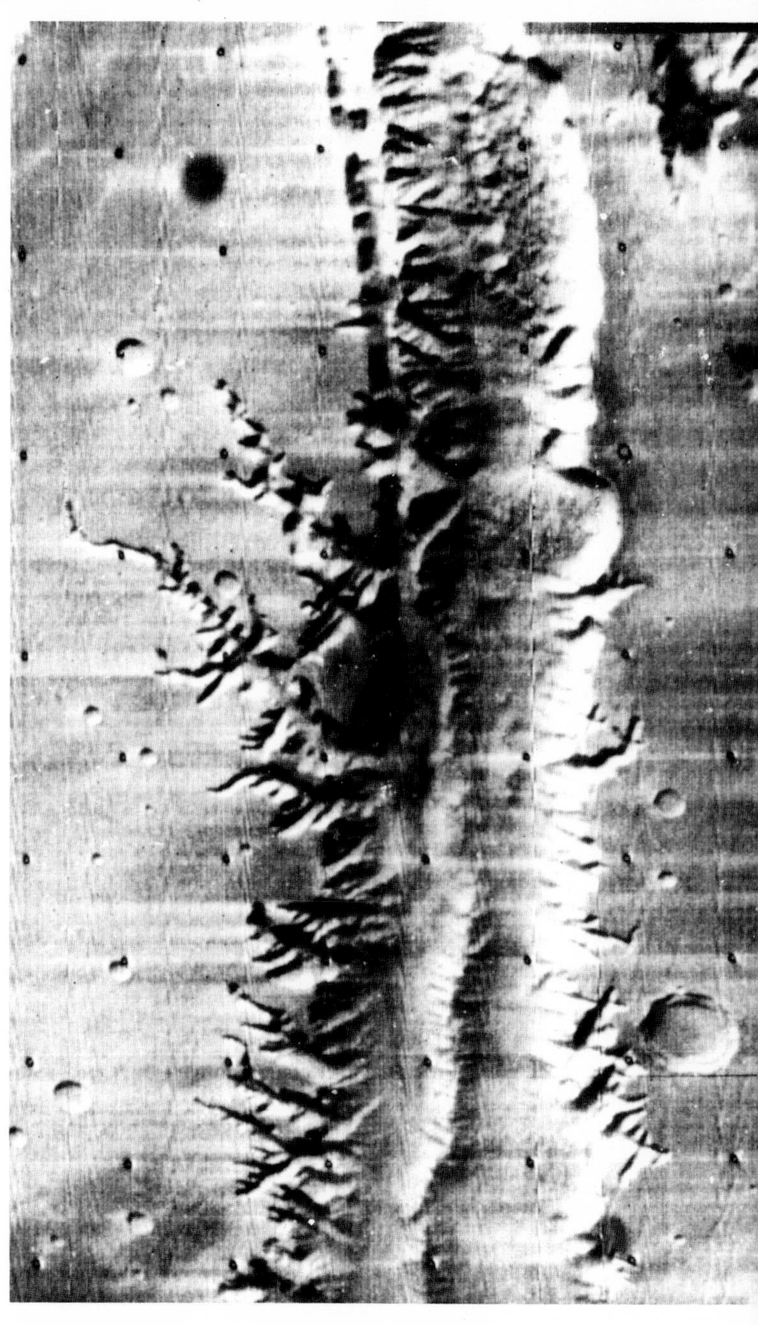

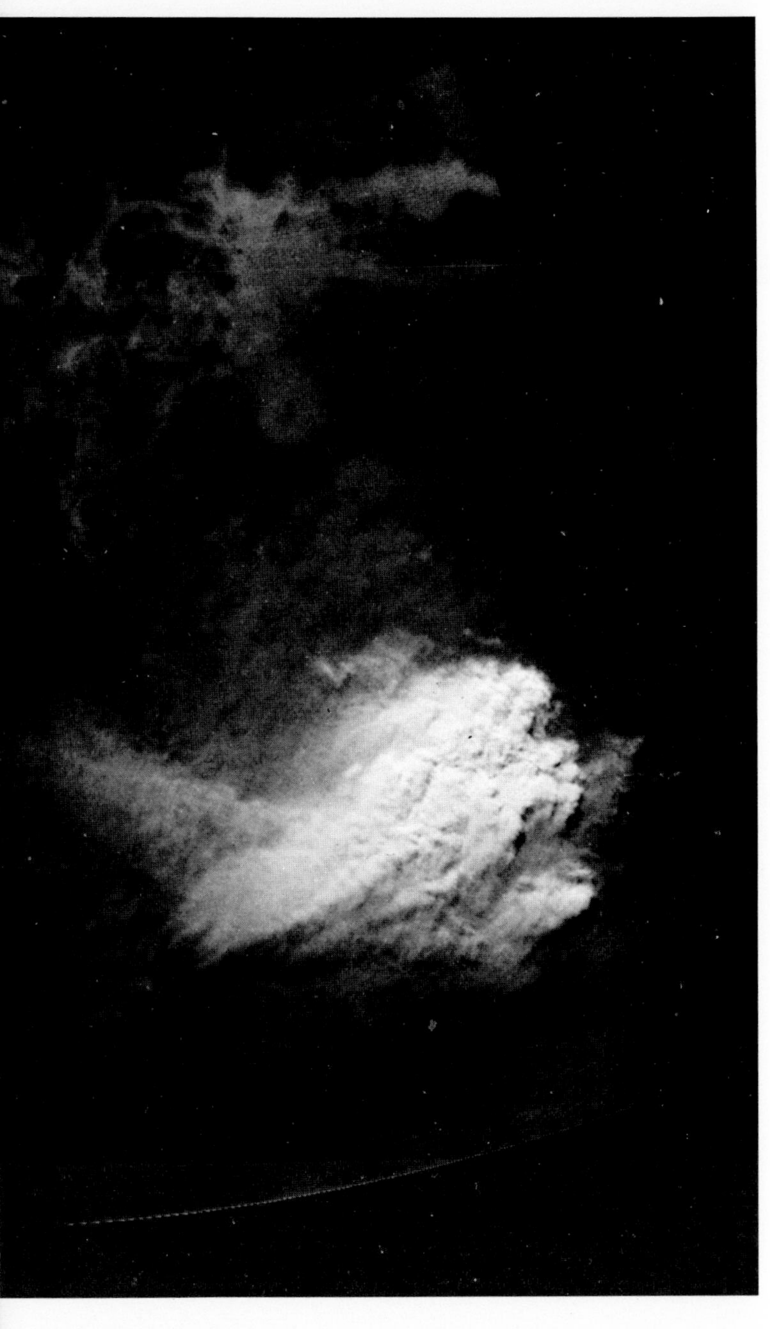

LEFT: This dust storm is the same as those on earth, but an order of magnitude larger. Even so, it is only a local disturbance by Martian standards. Once or twice a year a single unbroken storm wraps itself around the whole planet, smothering everything in a blanket of whirling dust. One of these storms occurred while the Viking mission was on its way to Mars and it seemed as if the project might have to be abandoned. Fortunately the dust clouds lifted a few days before the spacecraft arrived.

The curious shape of the picture is an effect common to all low-level orbiting spacecraft. The image is scanned a line at a time, starting at the top. By the time the scan reaches the bottom, the spacecraft has moved an appreciable distance and is scanning the surface further along its path. This results in a picture with two parallel sides and two curved ones. The picture may have to be "bent" still further to correct the parallax effect of the planet's curved surface.

FAR LEFT: The huge canyon of the Valles Marineris, which was named after the spacecraft, is on the equator of Mars and clearly visible through telescopes from the earth. It is a rift valley with a structure similar to the East African Rift. But if this canyon system were laid out on earth, it would span the entire United States.

Apollo astronauts, and tracks that could remain undisturbed for a million years.

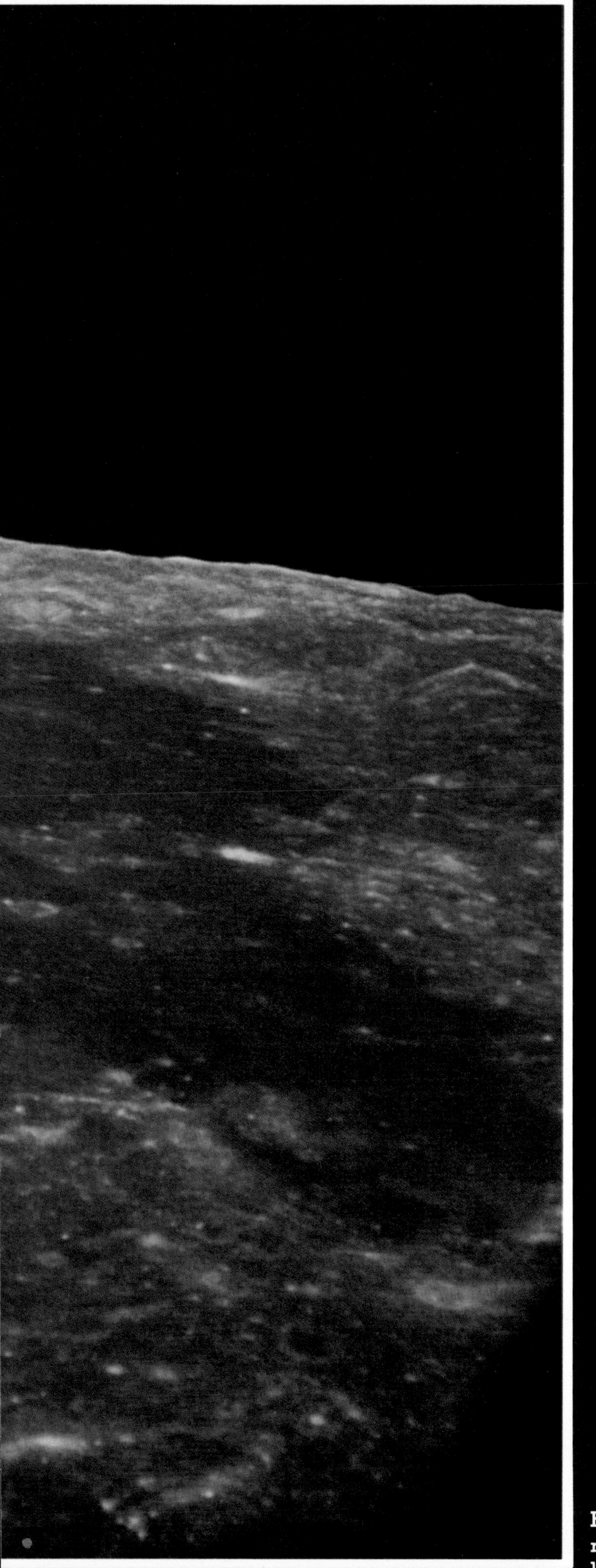

Earthrise, as the Apollo 11 lander makes rendezvous with its mothercraft for the return home.

THE EARTH

"We shall not cease from exploration
And the end of our exploring
Will be to arrive where we started
And know the place for the first time."
—T.S.Eliot

Of all the pictures we have received from space, the "whole-earth" image is the one instantly accepted as classic. The recognition that we are a part of this fragile isolated system gives it a strange emotional charge, a sense of identity with the planet, which could be one of the most important discoveries we have ever made.

A generation ago the idea would have been meaningless. In fact, the earth was the one planet astronomers had never seen as a planet. Yet 500 million people can now take a look at it from outer space every night by casually glancing at the TV weather forecast. Whole-earth pictures, like this Meteosat image (OPPOSITE), are almost a cliché.

The change in perspective this represents has been so complete (and natural) that we already take it for granted. But the familiarity is misleading. The exploration of earth from space has only just begun, and the view it gives us of ourselves – from the outside – is so unexpected it can be difficult to recognize.

With the familiar dimensions of up and down removed, the surface is two dimensional and most of our landmarks disappear. Mountains are flattened. Colors and textures appear which have nothing to do with our visible landscape. Lines are replaced by flow patterns of rock and water; and for the first time it is possible to see the earth as an ecosphere of systems interacting with each other at different scales and speeds.

The surveys carried out by orbiting satellites have produced an astonishing flood of images. The most prolific source is probably the NASA Landsat program, which produced more than 5 million pictures (based on some 200,000 scans) in the first 3½ years of operation, and now has an output of over 2 million a year.

If one adds the stream of data from dozens of weather satellites, the USSR's earth resources program, the Third World satellites and the European space projects, it means that in the last decade we have received between 50 and 100 million images of the earth from space.

And this is only part of the story, because not all space imagery is released. It is a sobering fact that two out of three satellites are military secrets, whose very existence is denied. But many of them are known to carry sophisticated imaging systems, so the figure could be doubled or even tripled. It amounts to an explosion of information which is almost beyond our capacity to use or understand.

An increasing number of organizations receive satellite data directly and reconstruct the images for themselves, but they do no more than sample the available flow and the techniques of interpretation are still in their infancy. A geologist will not see the same things in a picture as a town planner or a specialist in plant diseases, so each one has to develop their own ways of decoding them.

For the moment, not even the scientists who made them could tell you exactly what these images mean. We are all seeing them for the first time.

DECODING THE PICTURES

Earth satellites have certain advantages over deep space probes. Their scans are made from a distance of a few hundred, rather than a few hundred thousand, miles. The satellites are in position longer, so they can wait for ideal conditions and revisit an area many times; and since they have simpler guidance and control systems, they have room for more sophisticated camera equipment.

Many are fitted with banks of sensors called multispectral scanners (MSS), which can record a view in a number of frequencies simultaneously. This allows a wider range of options when it comes to processing them. By using different combinations of scans, it is possible to reconstruct a "realistic" image or a colored diagram of the same scene.

Landsat uses a four-way MSS, which operates in red and green visible light and in two bands of infrared. Each scanner, of course, produces a monochrome image. The traditional "false color" rendering is made by projecting each of the images through a different colored filter: infrared through red, red through green, and green through blue. When they are combined, water normally shows up as black, vegetation and other heat sources are red, cities and built-up areas are grey, and the average landscape runs through a range of browns, yellow and blue.

A variety of chemical and electronic processes can be used to achieve this, but these particular color transpositions have become part of the vocabulary of electronic imagery.

Fortunately for those who have to interpret them, earth satellite pictures have another advantage. It is possible to go to a place on the ground with a picture of it in your hand, and compare them. The key to this cross-match lies in identifying the "spectral signatures" of different surfaces. Stars are not the only objects that can be identified by their spectrum. Everything in the universe gives off its own distinctive pattern of radiation, whether it is generated by itself or simply reflected. Each signature is a unique combination of frequencies, which depends on the subject's reflective qualities, its chemical or crystalline structure, its heat, whether it is solid or liquid, how smooth its surface is and many other factors.

GROUND TRUTH

From a field of diseased wheat in central Asia to a battle tank in the Sinai Desert, everything has its spectral signature. If we knew all the spectral signatures, it would be possible to analyze the pictures in considerable detail. Geological processes could be deduced, mineral ores located and even species of plants identified.

Extensive surveys are now being carried out in different parts of the world to establish the "ground truth" of satellite pictures. NASA, for instance, has set aside certain target areas, which are scanned from space under different weather conditions and at every possible frequency. Teams of scientists go over these scans in meticulous detail, recording everything from the constitution of the soil to the level of atmospheric humidity.

The search for ground truth, like the list-making activity of any new science, is a laborious business. But we need to compile a dictionary before we can translate the language. What has already been achieved, as the following pages show, is remarkable.

LANDSAT

From the scars of strip-mining in West Maryland (TOP) to the tracks of nomads across the remote deserts of Libya (BOTTOM), the electronic scanners of Landsat keep watch on the world.

NASA's long-running earth resources program began with the launch of the first Landsat satellite (then called ERTS-1) in 1972. This was followed by a second satellite in 1975 and then by a third, each overlapping and eventually replacing its predecessor in order to maintain a continuous earth-watch.

Many of the pictures on the following pages are Landsat images, so it is worth looking at how they were taken.

For many years Landsat has crossed the Equator at exactly 9.38 A.M., fourteen times a day – an apparent impossibility caused by its near-polar, sun-synchronous orbit. Translated into simpler language, this means that the earth is spinning from west to east inside the steady circular orbit of the satellite over the poles, like the flywheel of a gyroscope. Each time the satellite crosses the Equator, the sun is in the same position in the sky (in other words, it is the same time of day), but the earth has turned beneath it to reveal a new piece of geography.

The crossing points move 99 miles (165km) along the Equator every day, so that the satellite actually travels along a series of parallel tracks, each scan overlapping the previous one by at least 10 percent. 252 orbits later the satellite finds itself back on the same spot. In this way Landsat can survey a continent-sized area in an eighteen-day cycle, though it usually takes two or three cycles to get a clear picture because of cloud cover.

From a height of 570 miles (912km) each complete scan covers about 13,225 square miles, with a resolution of 262 feet (80m) – that is, the scanners can distinguish objects 262 feet apart as separate features. This depends on the contrast, however. If they stand out strongly from the background, roads and bridges only 33 feet (10m) apart can be seen. It is possible, for instance, to make out sections of the Great Wall of China on Landsat pictures; and provided there is a change of vegetation or surface texture on either side, one can even detect the line of a narrow boundary fence.

As most of the orbits are over the sea, it would be wasteful to keep the scanners on all the time, so they are controlled from ground stations. There are three of these in the United States (Fairbanks, Alaska; Goldstone, Colorado; and Greenbelt, Maryland) and several in other countries including Canada, Brazil and Italy.

Small automatic devices can be set up in remote areas to turn on the satellite as it passes overhead. When it

comes within range of a ground station, it can be ordered to replay the scans it has recorded on the other side of the world or it can be given instructions for its next sweep.

EARTH-WATCH

The decision about which areas to survey depends on the growing list of clients who now use Landsat imagery – from governments compiling forecasts of annual crop yields to oil companies looking for a lucky strike.

Landsat made the geological surveys for the Aswan High Dam in Egypt, but next time it could just as easily be a project for the California State Forestry Department checking on forest fires or for ecologists monitoring the seasonal appearance of plankton off the North African coast.

The "guardian angel" factor – Landsat's value as an ecological watch-dog – was demonstrated soon after it went into operation when it detected the modification of weather patterns over a large area of Lake Michigan caused by the atmospheric pollution from steel mills in Gary, Indiana. It has since been used for pollution control of all kinds, and is especially good at detecting otherwise invisible discharges into coastal waters and estuaries.

As a means of mapping remote areas, earth satellites have proved to be 95 percent as accurate, and 95 percent cheaper, than conventional ground surveys. In fact, the comparative costs of ground surveys, high-altitude aircraft and satellite imagery have been estimated at $50, $15 and $1 a square mile respectively.

The importance of this can hardly be overstressed. In the past, poor countries could not even afford to find out what resources they possessed, let alone make use of them. When the resources include such basic elements as water, it can be a matter of life or death.

Anyone who thinks they are familiar with the world may be surprised to learn that 10 percent of the land mass is still totally unexplored and that vast areas of Asia, Africa and South America are poorly or inaccurately mapped.

Landsat has already recorded unknown rivers and mountains, not to mention a large new meteorite crater in Brazil's Mato Grosso. It has even made discoveries in downtown USA. In 1974, for instance, it revealed an entire lake, just north of Houston, Texas, which was not shown on existing maps!

THE WORLD'S WEATHER

The earth's atmosphere is a lucky fluke. We are the only planet with liquid water on the surface; at exactly the right temperature to keep it evaporating and condensing (without ever boiling off); with the precise mixture of gases to shield us from radiation, yet allow the heat to escape; and with a varied enough surface (thanks to the seas and oceans) to produce the intricate, interlocking patterns of our weather.

The whole system balances on a knife edge. If the average temperature fell 10°C, there would be a Martian ice age; if it rose 10°C, the Venusian greenhouse effect would take over.

Yet major changes have certainly occurred in the past, and the system is imperceptibly altering all the time. The angle of the earth's axis, for instance, varies over a 40,000-year cycle, and there is an irregular wobble that alters the seasons – the "precession of equinoxes" – every 20,000 years. Ice ages come and go at 100,000-year intervals, probably caused by variations in the sun's output and other factors, such as close encounters with comets.

Nor are we immune to events outside the solar system, such as the forty or fifty supernovae that have exploded in our region of the galaxy during the earth's history.

The weather has so many variables that no one has yet produced a successful model of how it works. The limits of accurate weather forecasting used to be about thirty-six hours. Computers and space technology have extended this, with difficulty, by about a day, but the ability to predict a full week ahead is still no more than a fantasy.

Unfortunately, our ability to sabotage the system far exceeds our understanding of it. Recent research shows that the steady build-up of sulphur compounds and the erosion of the atmospheric shield by fluorocarbons are both reaching potentially dangerous levels. So the next major change will probably be caused by our own pollution.

Ironically, it is not the first time this has happened. It was the exhaust gases from the early life forms (a scum of algae floating in poisonous seas) that first added oxygen to the atmosphere and made the planet habitable 3.5 billion years ago.

BELOW: **The world's weather for May 2, 1977, shown in a mosaic of images from the NOAA low-level satellite around the Equator.**

The picture was taken over a twenty-four-hour period, so the result is like a lengthy time exposure. But events such as the hurricane nudging Madagascar and the curtain of cloud reaching back over central Asia from the Himalayas can be seen in meticulous detail.

Large-scale meteorological systems are also apparent. For instant, the equatorial projection graphically illustrates the Coriolis Effect, whereby the northern and southern weather systems are set spinning in opposite directions because the earth's surface is moving faster across the middle of the picture (from left to right) than it is at the top or bottom.

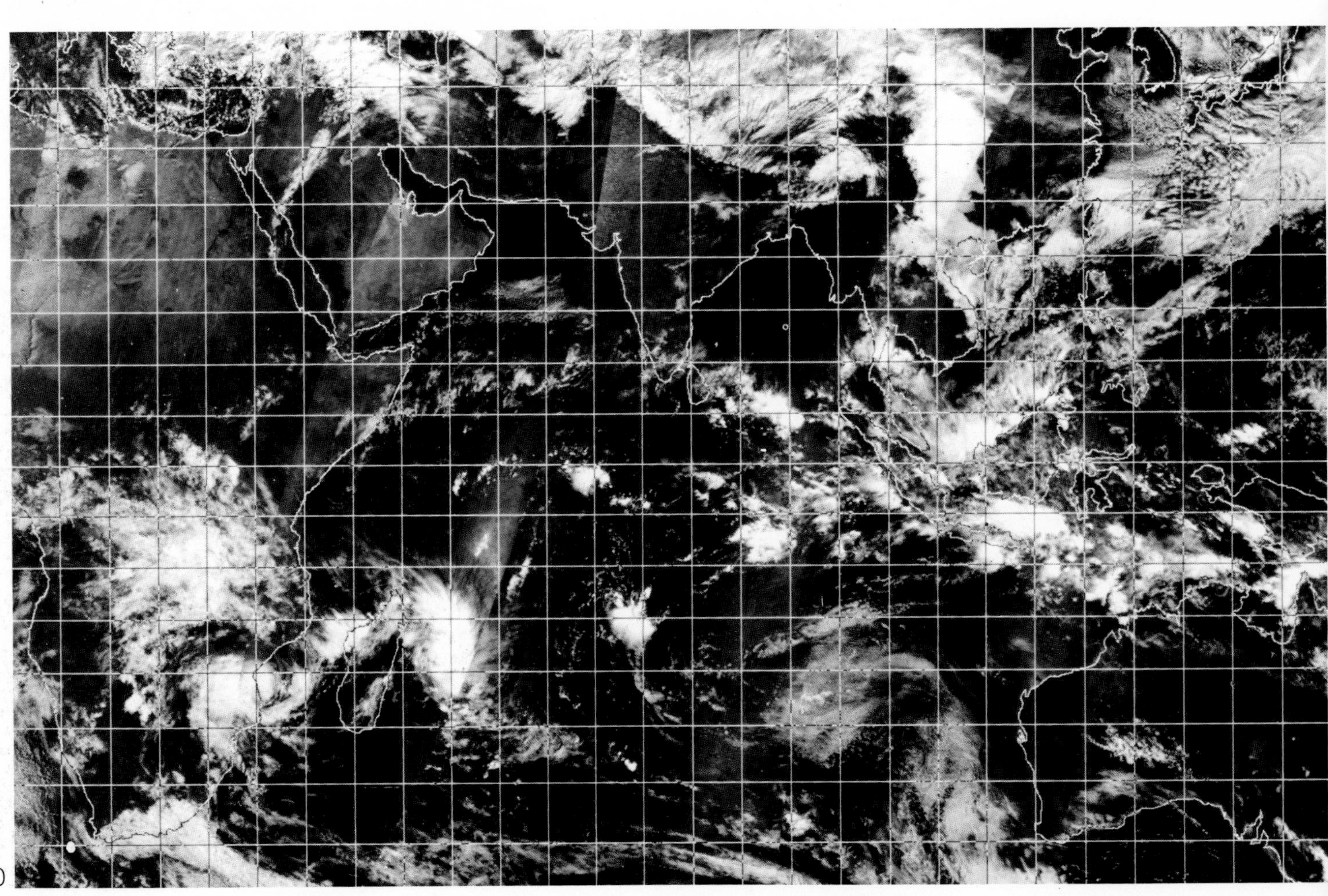

METEOSAT

Meteosat is a European Space Agency satellite which was launched in late 1977 into a geosynchronous orbit over the Equator. Turning with the earth, it holds its station on the Greenwich Meridian (0° longitude), while its three-way scanning system produces high-resolution images of Africa, Europe and the Middle East.

The European space effort is now gathering momentum. Satellites carrying ultraviolet and X-ray telescopes are in operation, further Meteosats are planned, and European astronauts are now in training for the ESA orbiting laboratory, Spacelab.

Weather forecasters need to know more than the shape of clouds. There is no way as yet of measuring barometric pressure from outer space, but Meteosat pictures like these of heat and moisture patterns can be combined with images of the same scene taken in visible light to produce a composite image.

TOP: **A picture of the heat reflected off the earth at infrared wavelengths. As infrared rays are easily absorbed by atmospheric moisture, hot dry areas like the Sahara and the Middle East are sharply defined, while tropical Africa is invisible because it is so humid. The heights of clouds, which are indistinguishable in ordinary photographs, show up as different shades of gray. The darker, lower ones contain more moisture; white indicates the cold tops of high clouds.**

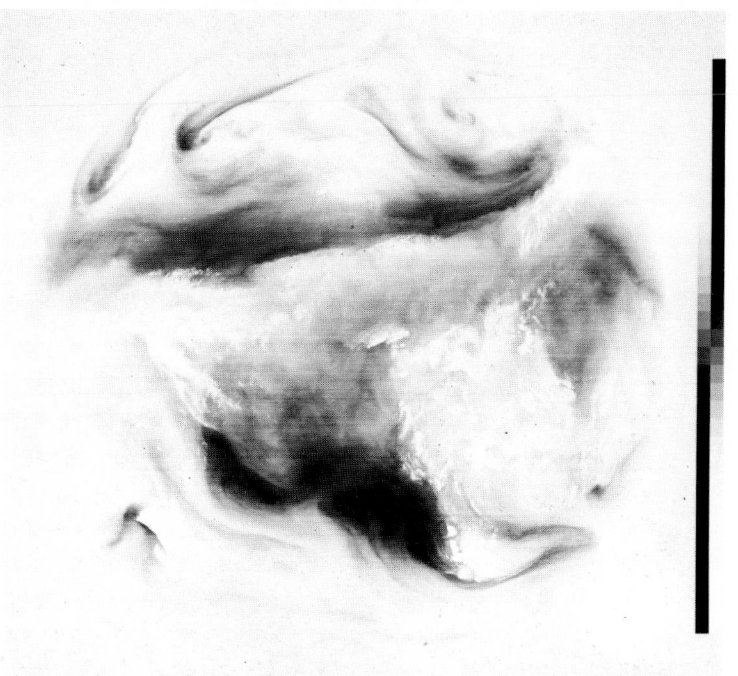

CENTER: **The same view as above, but this time the image was made at the wavelength most easily absorbed by moisture. Dry is dark, wet is white, and the surface has disappeared under what looks like unbroken cloud cover. In fact, the white areas indicate high humidity between 3 and 6 miles (5 and 10km) above the surface.**

BOTTOM: **The only one of these pictures taken at visible wavelengths, where light and shade mean what we expect them to. The definition is sharper because of the widely different ways that surface materials reflect light.**

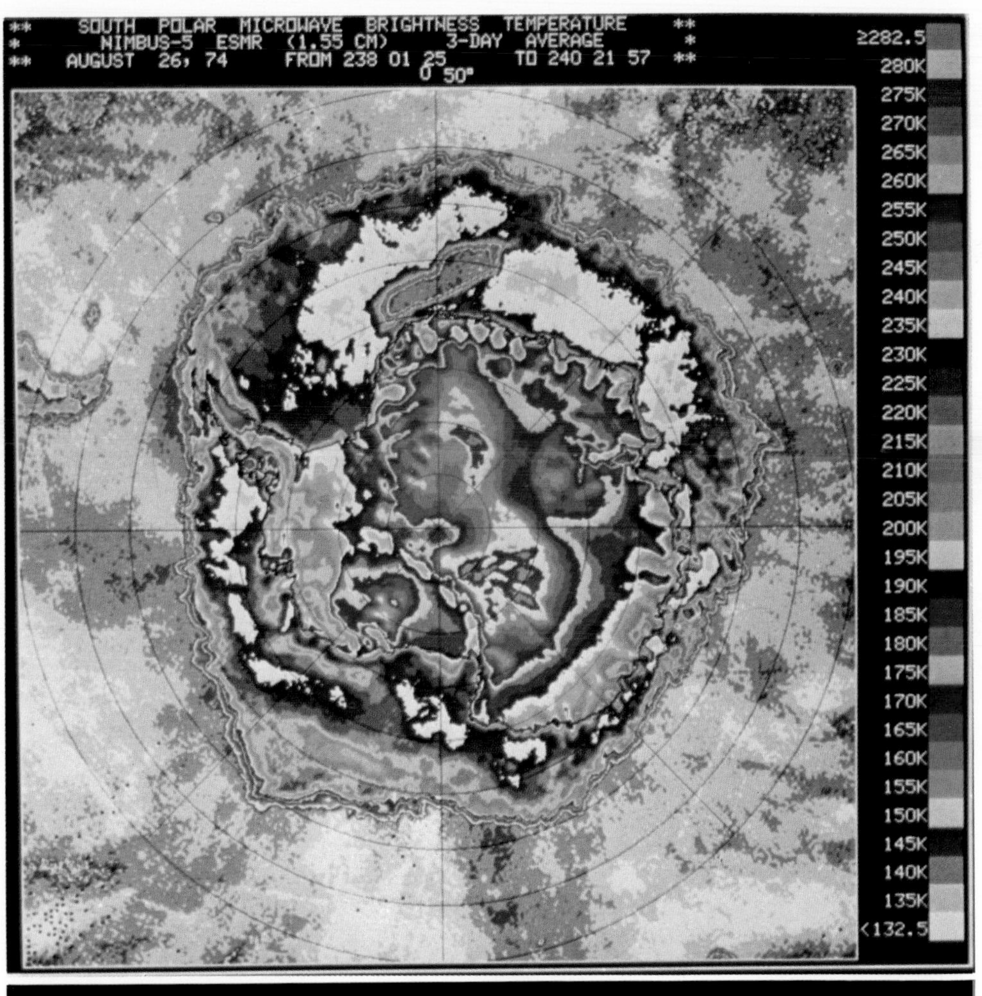

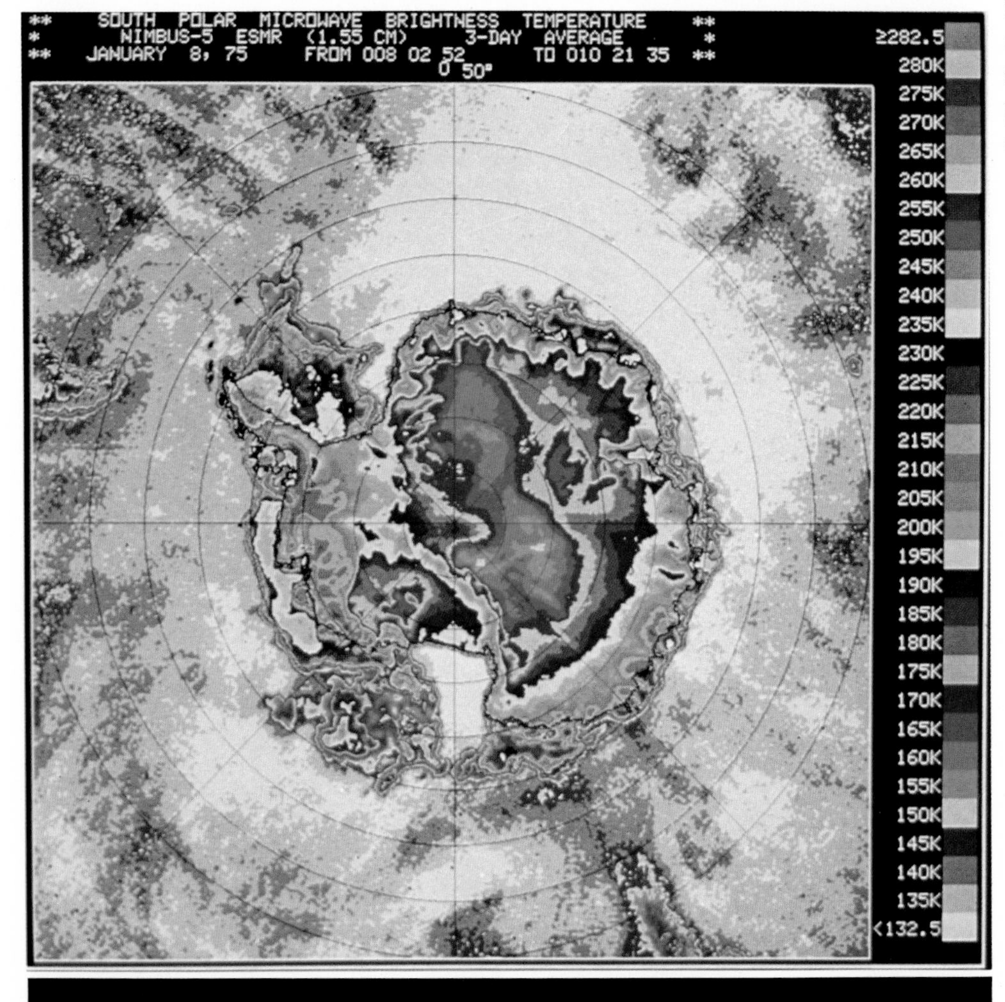

THE POLAR ICE SCANS

The whole of Antarctica is revealed for the first time by wide-angle instruments on board the Nimbus 5 weather satellite.

It is winter and from a viewpoint over the South Pole (TOP LEFT) it is possible to see the scale of the ice sheet that protects the last great unexplored landmass on earth. The image bears little resemblance to a conventional photograph because it records a characteristic called "brightness temperature." Brightness temperature is the amount of radiation (in this case 1.55cm microwaves) given off by an object, multiplied by its surface temperature. Some surfaces radiate better than others, so that although the sea water is warmer than the ice shelf, the ice emits more microwaves and has a higher brightness temperature. On the other hand, the snow covering the inland plateau is roughly the same temperature as the ice, but since it is only a moderate radiator it has a cooler reading.

During the brief Antarctic summer (BOTTOM LEFT) the great ice sheets melt away. Maps and atlases will have to be revised because these pictures indicate much less pack ice than is usually shown. The shape of the ice shelf is also more complex than was previously thought, with irregularities caused by the vast circumpolar current that endlessly circulates around the continent in a clockwise direction, acting as a regulator for much of the world's climate.

A winter view of the North Pole (TOP RIGHT) and the frozen wastes of the Arctic Ocean which, together with the South Polar ice, covers 12 percent of the world's seas.

For the first time scientists can begin detailed studies of these areas because microwave radiation can be measured through clouds (unless it is actually raining) and shows up details of ice structure which are no more than a white glare on ordinary photographs. Freshly formed ice can be distinguished from old, snow cover measured and icebergs tracked. The limits of the winter freeze-up can now be recorded year by year, so that man-made effects on the climate or the onset of the next Ice Age can be detected as soon as changes occur.

Summer in the Arctic (BOTTOM RIGHT), but the almost land-locked ocean remains frozen over. From 690 miles, (1100km) above the North Pole, the electronic eyes of Nimbus 5 reveal the dramatic effects of the Gulf Stream as it sweeps warm water past Scotland and Norway and along the coast of Russia, while far to the south there are parts of Greenland which remain locked in year-round pack ice because of a cold current flowing down the eastern coast.

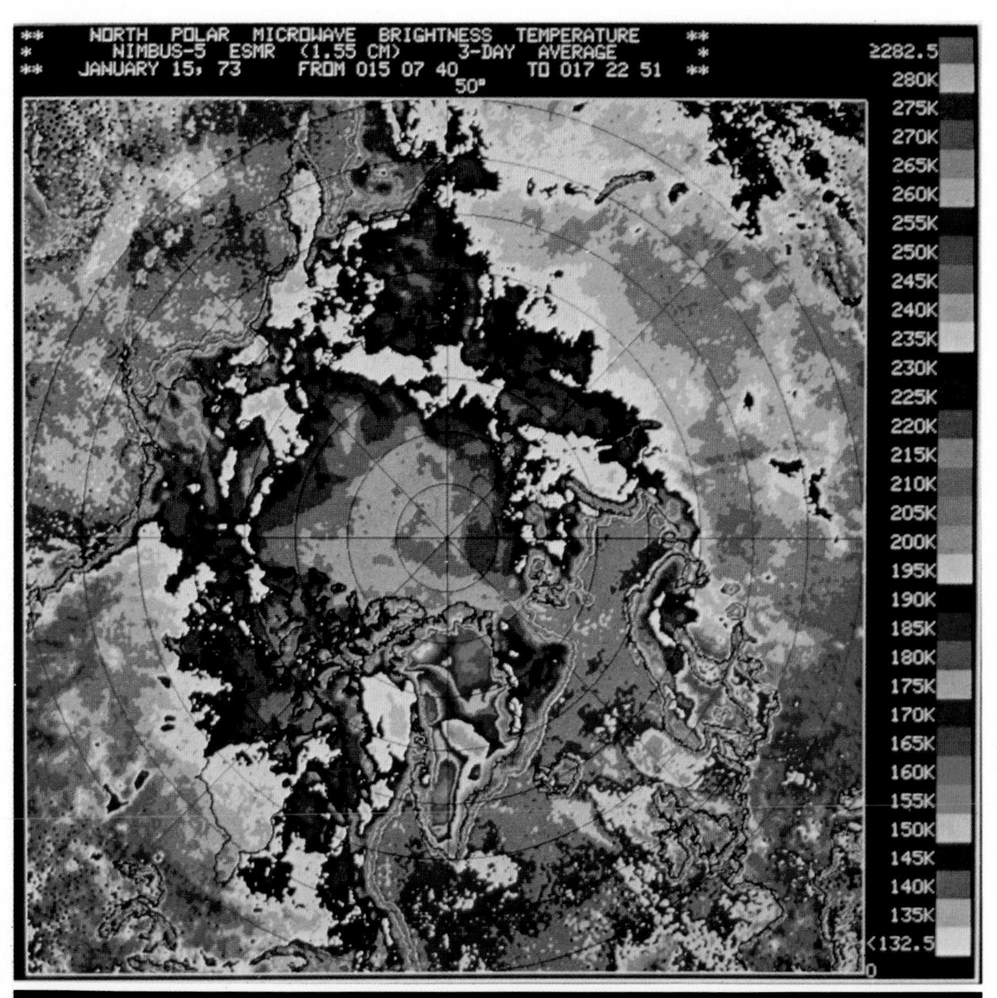

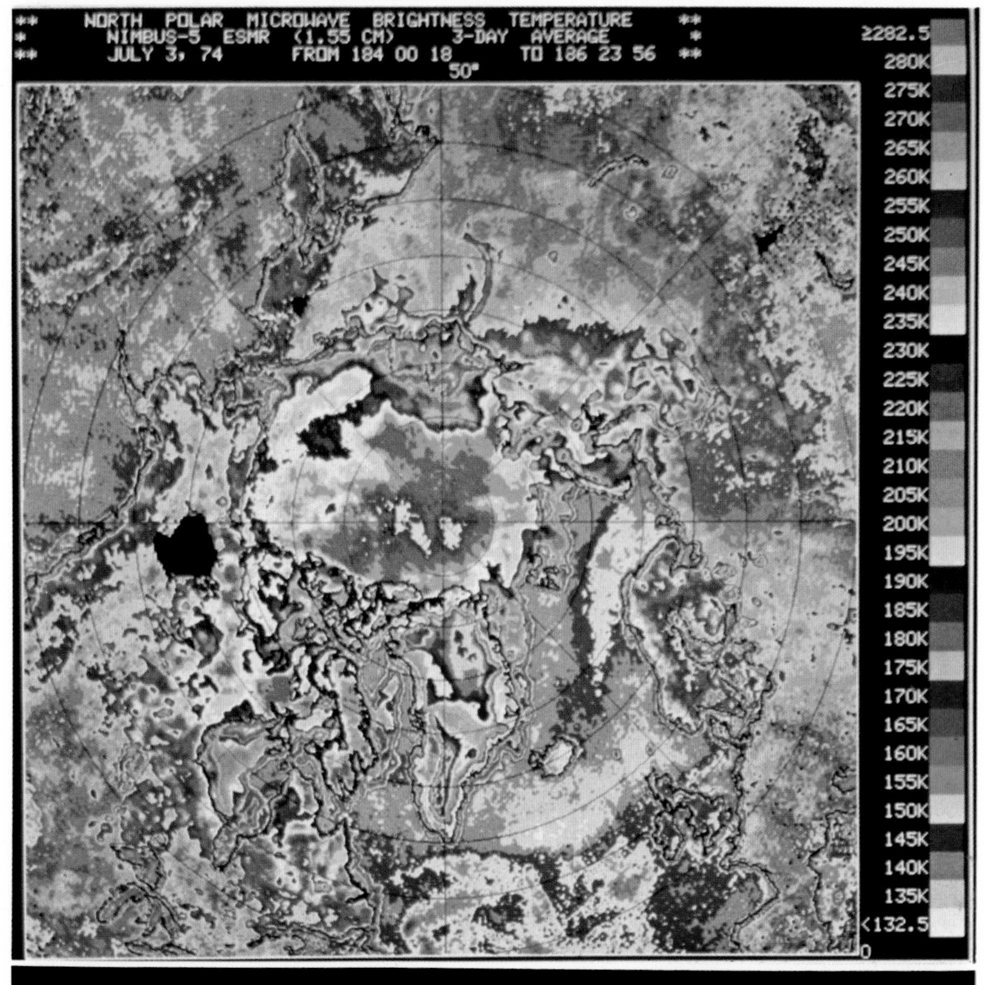

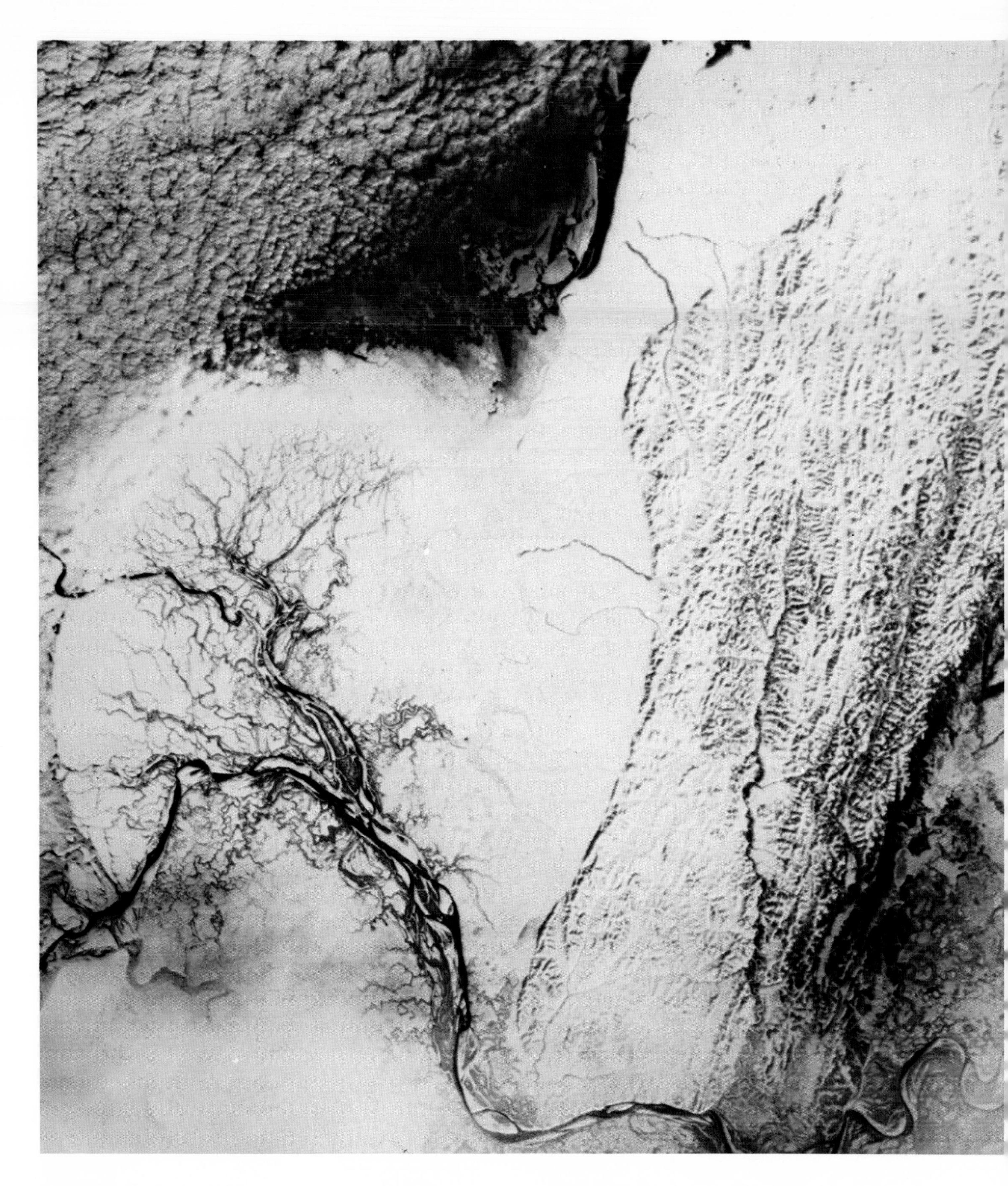

ABOVE: **A Landsat picture of the Yukon River in Alaska, at the point where it flows into the Bering Sea. The perfect semicircle of the delta is partly obscured by the sea ice that has formed along the coast, and through which the river has cut its own meandering path. Every minor channel shows up under the light dusting of snow along the coastal plain, a phenomenon especially useful to satellite photography because it reveals human artifacts such as field boundaries.**

OPPOSITE: **Sea ice can take delicate as well as threatening forms, as this view of Bonavista Bay, Newfoundland, shows. The blue coils swirling past Cape Bonavista like cigarette smoke are a mush of half-melted ice and small floes called brash, which is driven along the coast by the wind and currents. So accurately do they reflect the forces acting on them, that they can be read as a slow-motion meteorological map.**

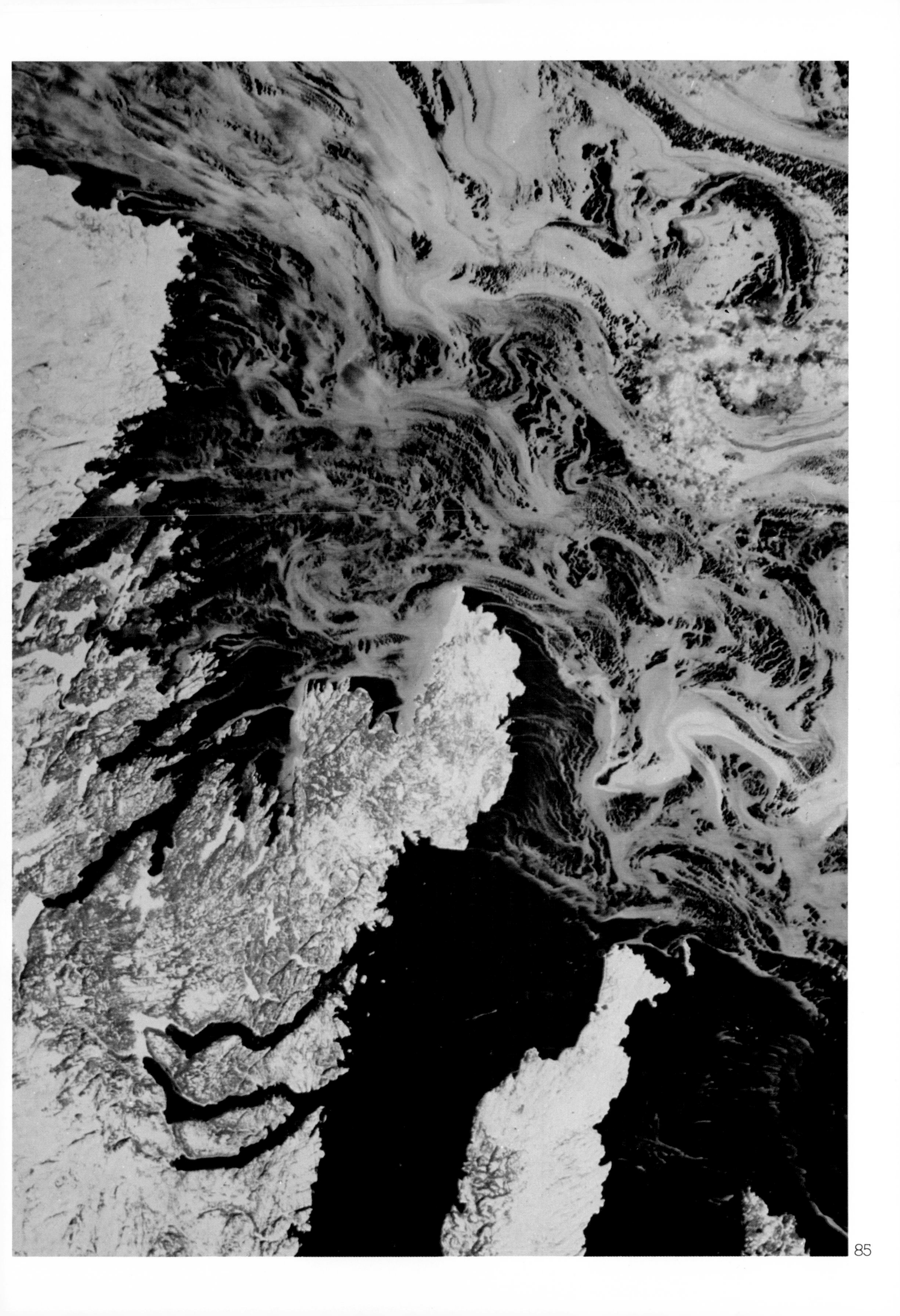

SIBERIA
The river Viljui meanders across the permafrost zone of the Russian Arctic. Although very little of its space photography is released by the USSR, this photograph taken from the Soyuz 22 spacecraft shows the extraordinary high standard the Russians achieve.

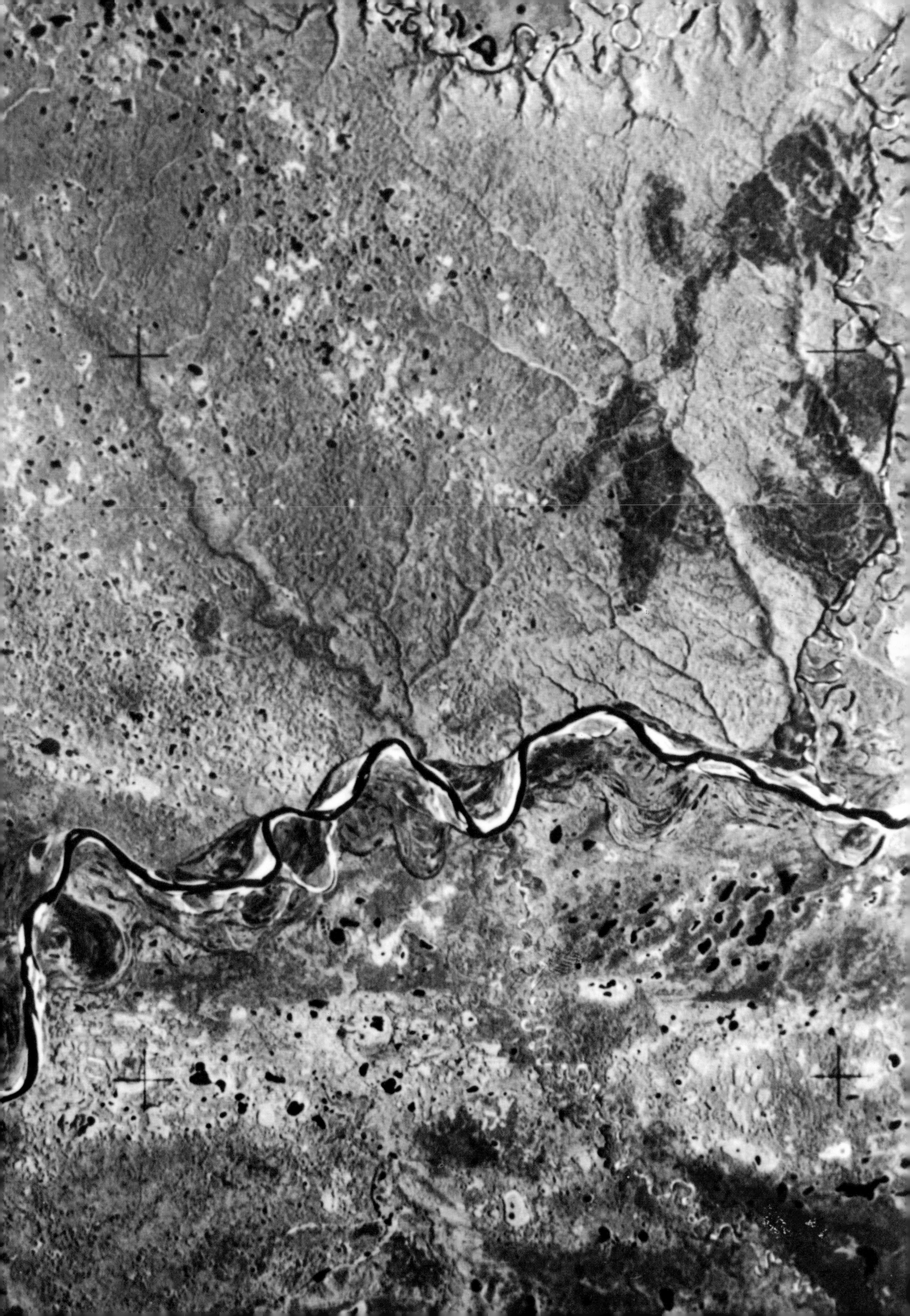

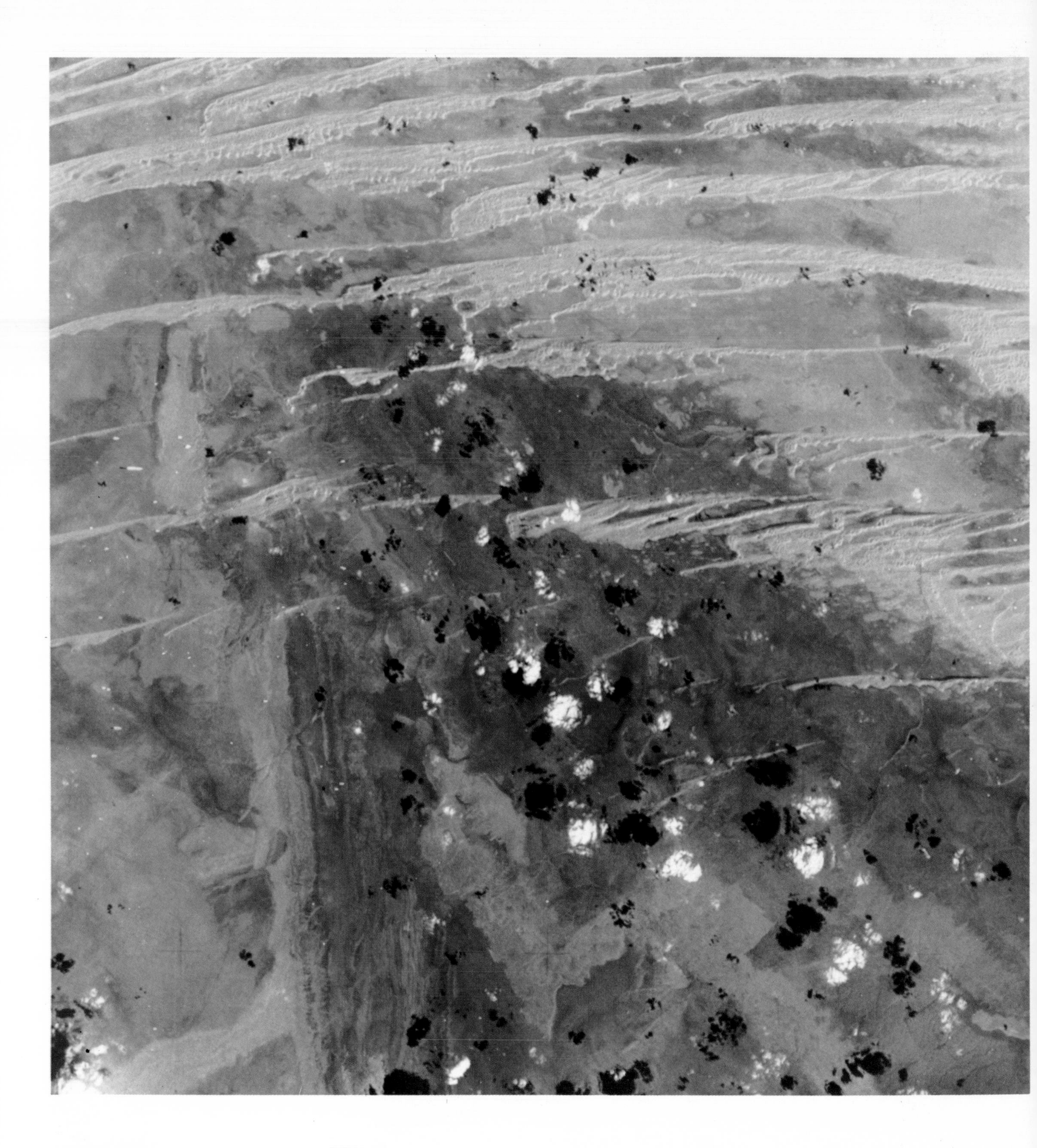

DESERTS

ABOVE: **This blue and yellow abstract, like a paint-spattered wall, is the archetypal desert, the Sahara. In the vocabulary of electronic imaging all deserts are blue. Whether they are composed of snow, rock, sand or ice, they have one characteristic in common – the lack of infrared radiation from living organisms.**

The streamers across the top of the picture are lines of sand dunes, often hundreds of miles long. The black spots are the shadows of the white spots, which are clouds. The small orange specks (left) are flares burning off the surplus gas from oil wells.

OPPOSITE: **The golden filigree of dunes at Rub'al Khali, deep in the central desert of Saudi Arabia, covers one of the world's great natural resources. When Arabia broke away from Africa, millions of years ago, the center of the peninsula buckled downwards into the fold which now contains some of the largest known oil deposits. The basin later filled with sediment from around the rim and formed a crust of salty mud called sabkhah (blue background). The curious patterns of sand are barchan dunes, cusp-like pockets a mile or so across formed by prevailing north winds (top left).**

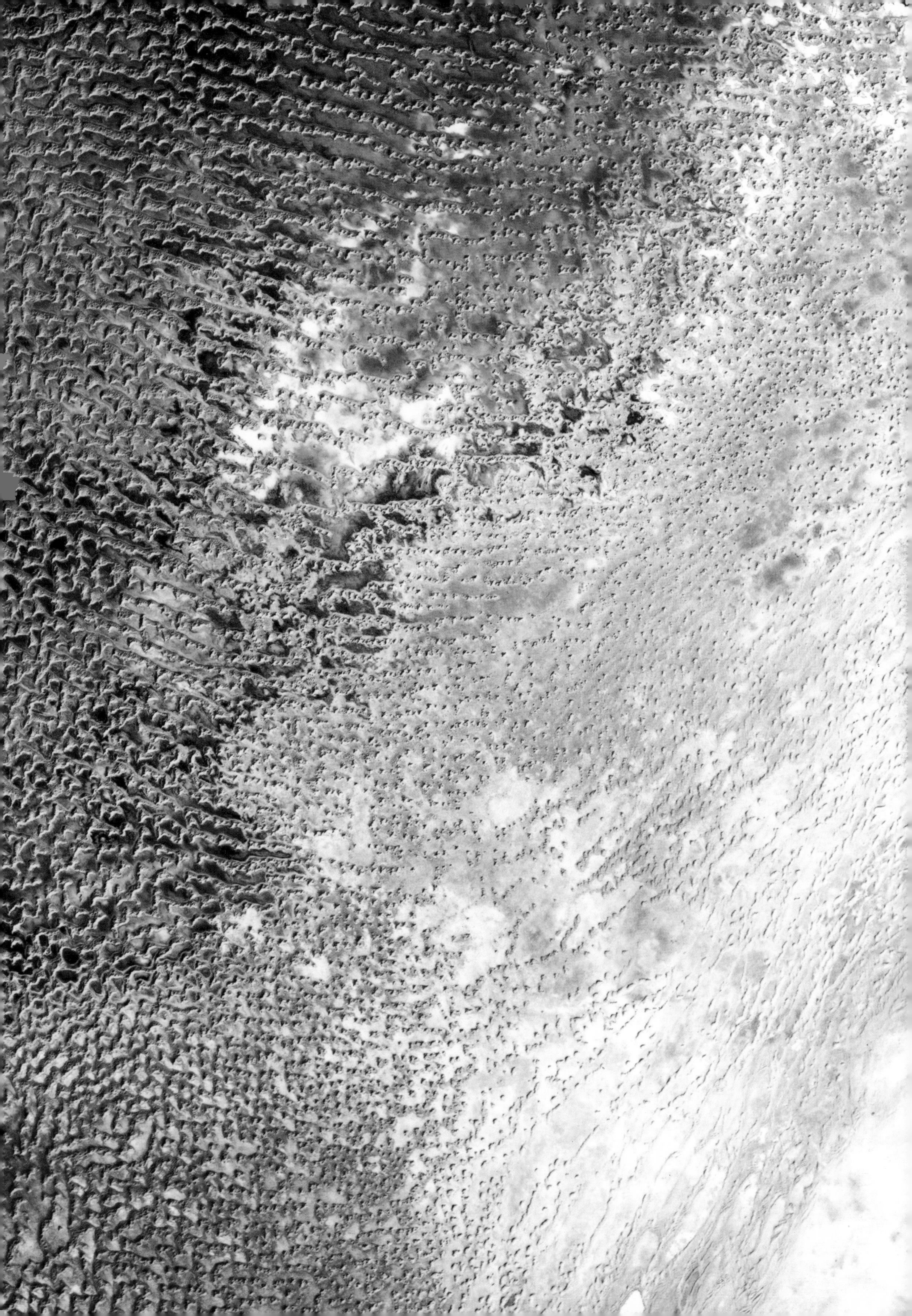

ABOVE: **From a height of 570 miles (912km) Landsat looks down on the uninhabited center of Algeria. The golden braids of sand dunes (top) are part of a desert called the Great Eastern Erg. The drainage patterns running out onto the gravel plain (below) show that the Sahara was once as well watered as Mars, a fertile plain flowing with rivers which became a runaway dust bowl.**

OPPOSITE: **The geological turmoil of southern Iran, shown in a composite of twelve Landsat images. Composites like this allow for startling definition and when they are coded with the natural colors of the visible spectrum, they have a super-realism. At the bottom of the picture are the busy waters of the Persian Gulf. The 14,000 foot (4,200m) mountains are in one of the world's major earthquake zones. The violence of the geology is indicated by the eruption of massive salt domes forming off-shore islands.**

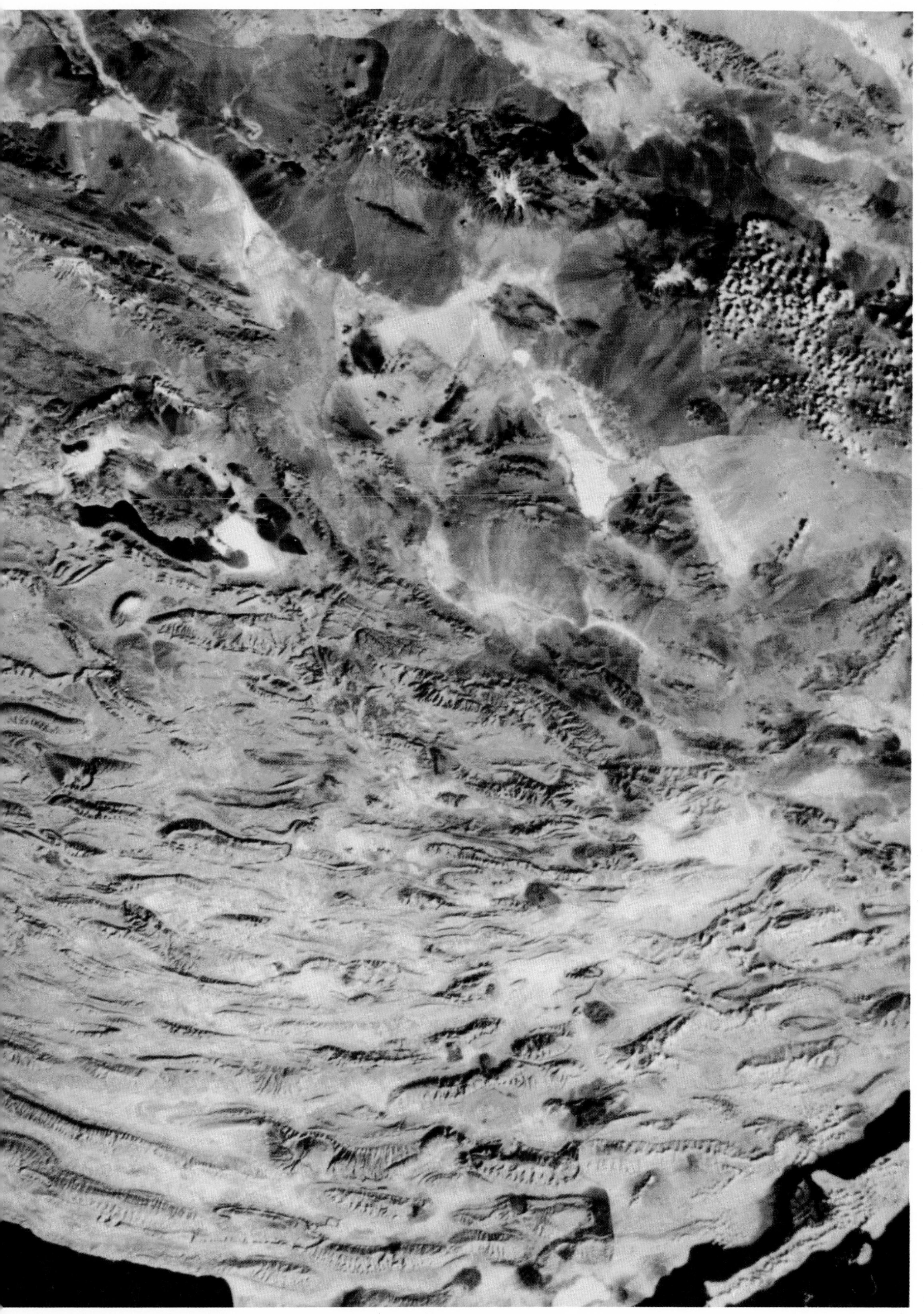

BALUCHISTAN

The mountainous frontiers of Islam meet in the Baluchistan desert.

The river Lahlab, flowing across the center of the picture, marks the boundary between Iran (below) and Pakistan (above), while the sand dunes (top right) are part of Afghanistan.

The burnished colors are no accident. The mountains are rich in metal ores, including deposits of porphyry copper at Saindak (top left corner), and the signatures of the Saindak deposits have been used to "train" computers to recognize similar reflectance characteristics elsewhere. More than twenty possible deposits have now been identified in this one picture and five of the most promising are being investigated by Pakistani geologists.

Few of the pictures in this book are credited to individuals because they are usually the result of teamwork, but this experimental work by R. Bernstein of IBM is an exception. Though it was made in the early years of the Landsat program, it sets a standard that has rarely been equaled by later, more sophisticated techniques. The quality of information is so high and the resolution so detailed that it is difficult to believe it was reconstructed from nothing but a string of numbers on computer tape, without even the use of a preliminary negative.

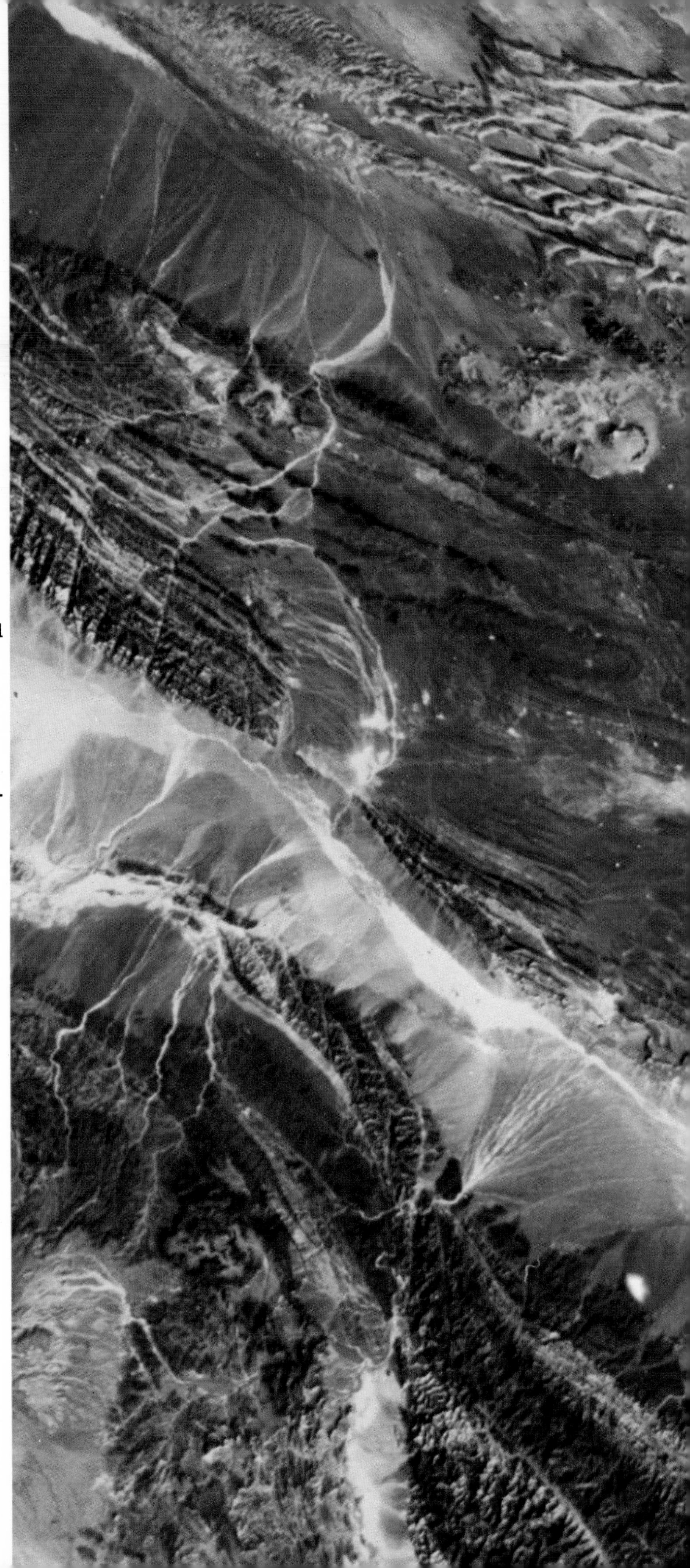

ABOVE: **Another desert on another continent, but this time a heavily populated one. The town (left center) is Las Vegas, Nevada, and the river is the Colorado. Las Vegas is located on the bed of one of the many dried salt lakes in the area. The "Strip" with its casinos runs roughly north-south. Both the lakes in the picture are artificial. Lake Mead (top right) was created by the spectacular Hoover Dam. Lake Mojave was formed by the Davis Dam further downstream.**

OPPOSITE: **A narrow strip of cultivation lines the banks of the Rio Grande as it winds through the basalt desert of New Mexico, with the town of Albuquerque (top right). The black areas on the left are beds of lava exposed by erosion and an extinct volcano is visible in the northwest corner. The dark areas at the bottom of the picture are shadows cast on the surface by clouds and the contrails of two aircraft which were passing at the time.**

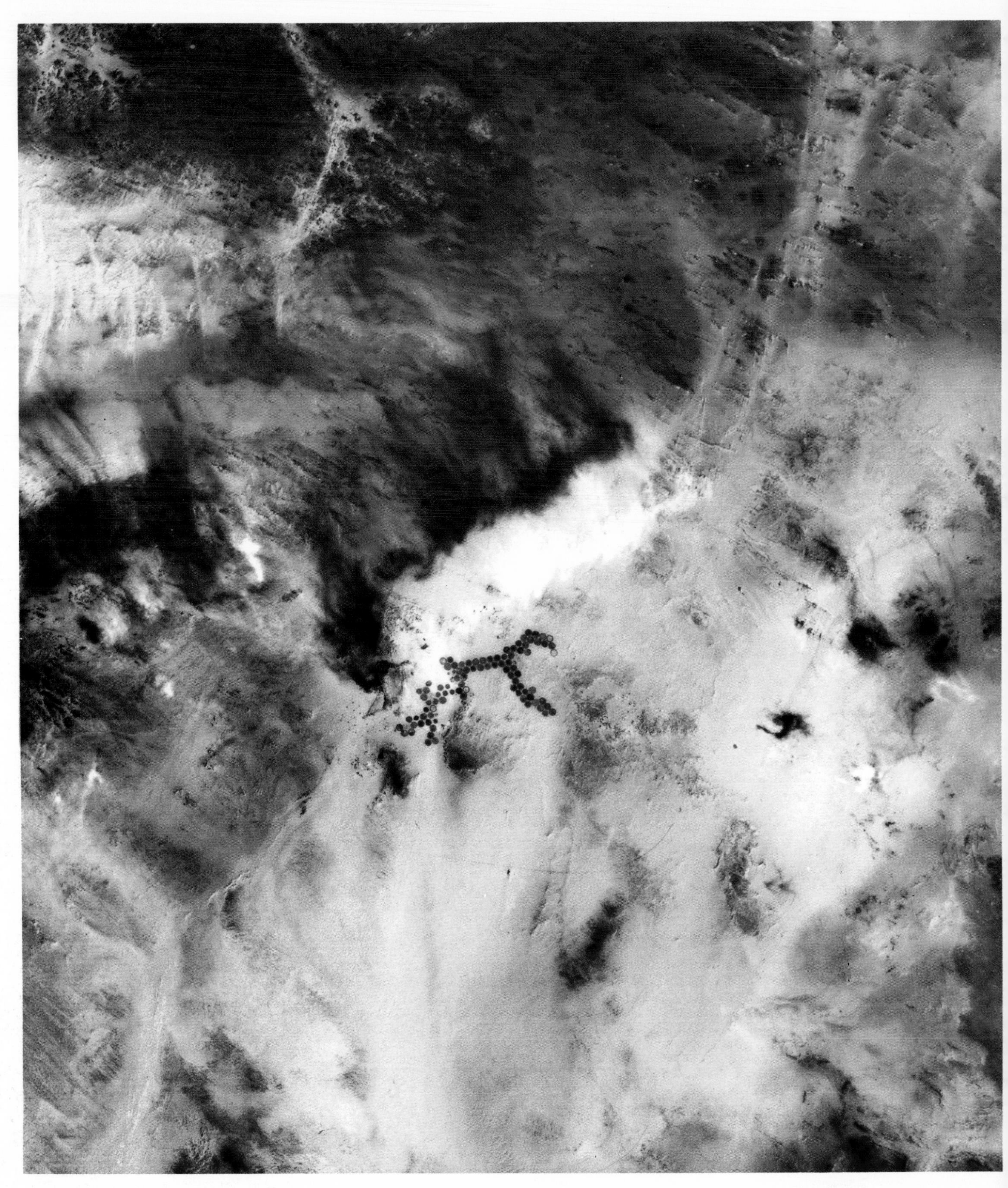

ABOVE: **A desert reclamation scheme in Libya stands out like a colony of red bacteria against the sand. Each tiny hexagonal shape is an artificial oasis, surrounded by a protective embankment and watered by rotating mechanical arms. There are no signs of irrigation ditches because the scheme is supplied by "fossil" water, discovered in deep rock strata and pumped directly to the surface. In desert areas it is as valuable as oil and, like oil, it is a limited resource that cannot be renewed when the supply runs out. There are thought to be similar, huge "waterfields" under the Sahara which could one day be used to bring the desert back to life.**

OPPOSITE: **Rivers of white sand snake through the mountains of Saudi Arabia. The coastline faces the Sinai peninsula, across the northern end of the Red Sea. The dark brown area (top right) is a prominent lava flow, and the orange-colored rocks are sandstone.**

The Nile Delta, and the field pattern of the world's oldest agriculture are outlined in vivid geometry against the desert.

HAWAII
The island of Hawaii described in two visual languages. The chemical image (ABOVE) was photographed by a Skylab crew-member in 1974 using a hand-held Hasselblad. The electronic image (BELOW) was reconstructed from Landsat numerical data and color-coded to enhance the lava flows down the side of Mauna Loa. The pastel colors are an example of electronic imagery flexing its potential. At a practical level, by stretching the tones as well as the hues of colors, the range of the code is widened and it is easier to exaggerate contours. Notice how vividly the lava flows stand out against the pink of vegetation and the yellow and white clouds.

Pictures like this show a curious relationship that has emerged between art and science. At first colors were chosen more or less at random, but it has been found that, when it comes to interpreting the pictures, researchers subconciously reject ugly colors and spend more time on attractive images. So, in effect, a "beautiful" picture actually reveals more information.

If electronic imagery can be thought of as a language, then this sort of relationship (between aesthetics and information flow) could be the beginning of its "grammar."

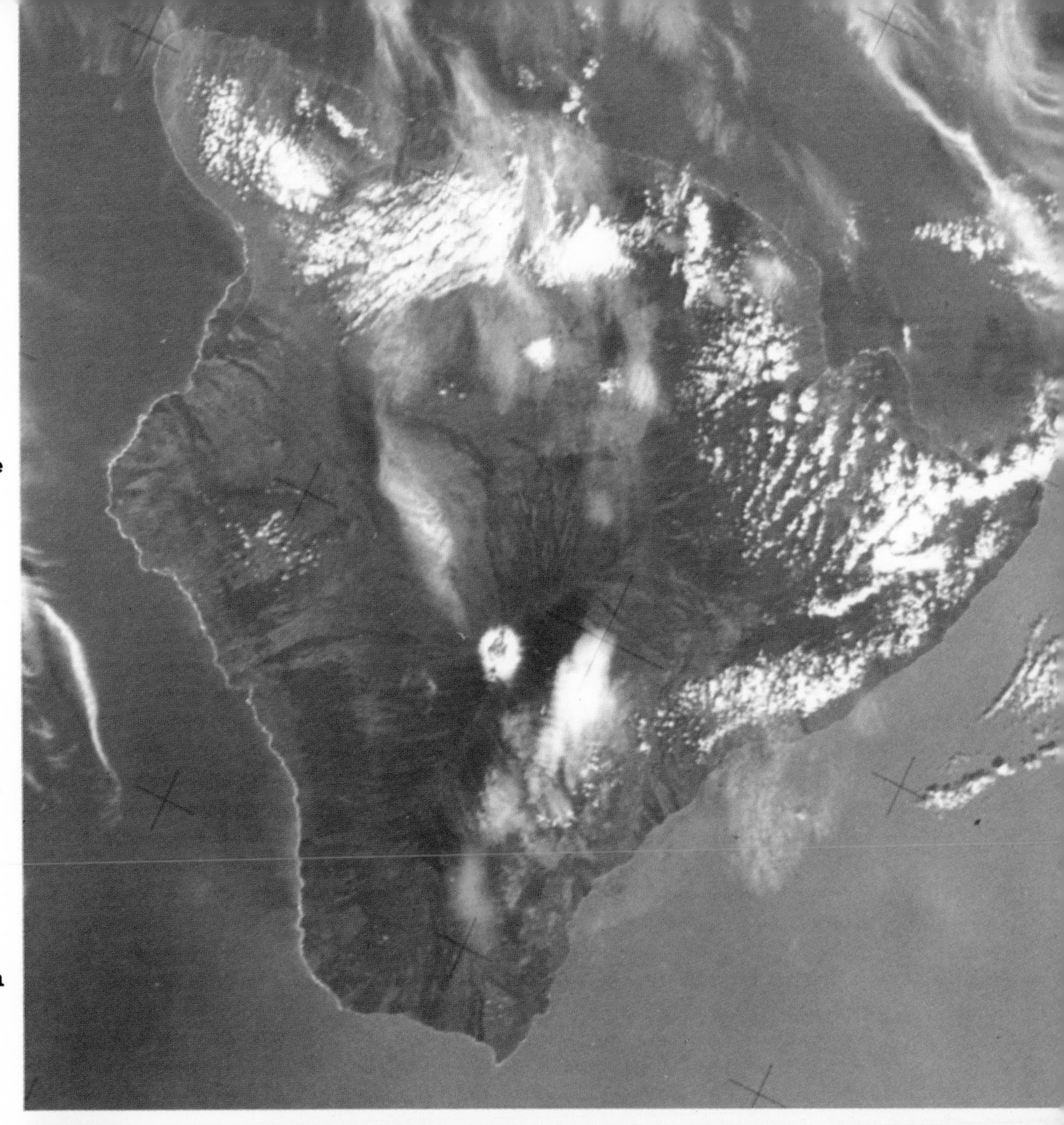

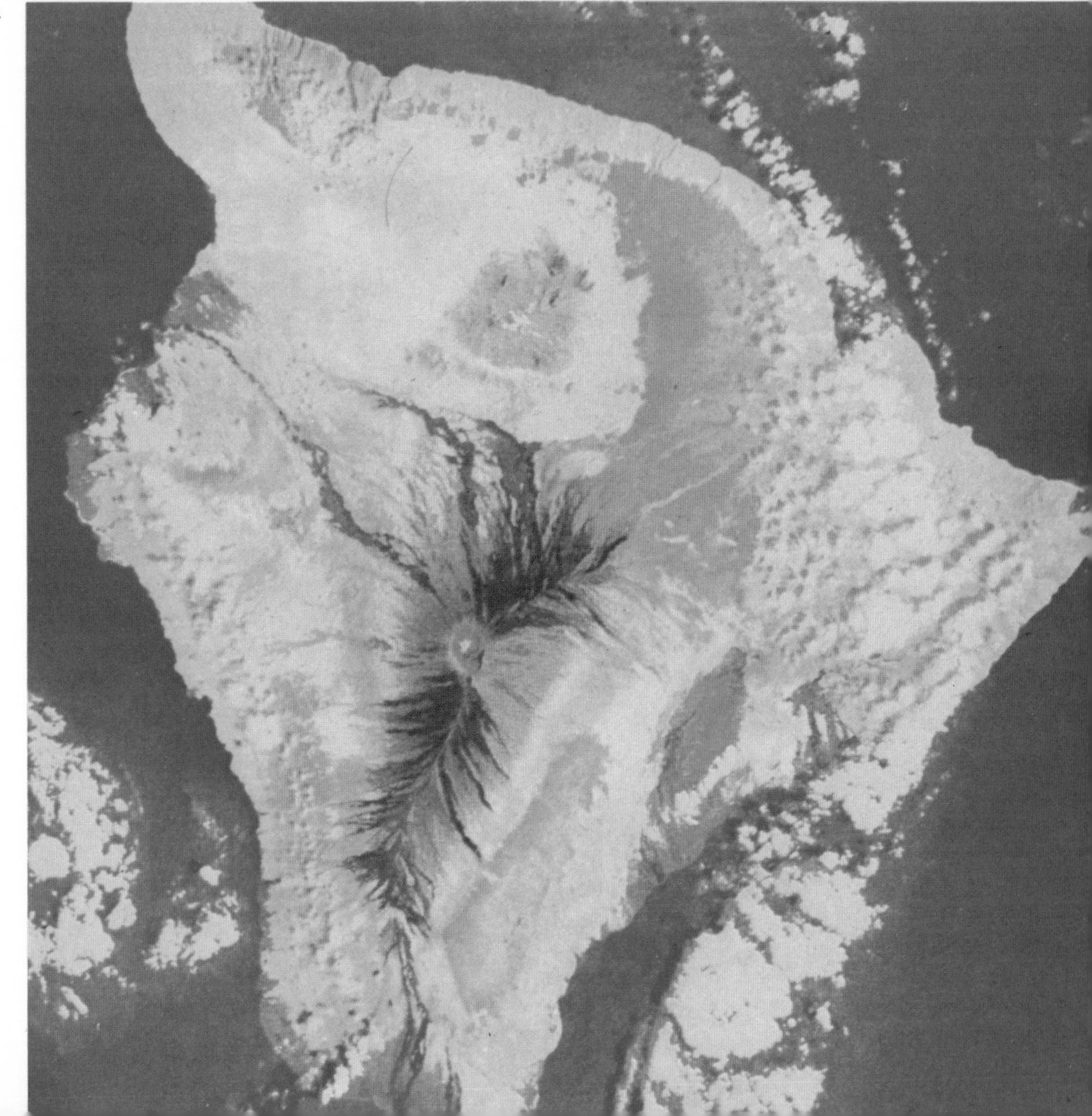

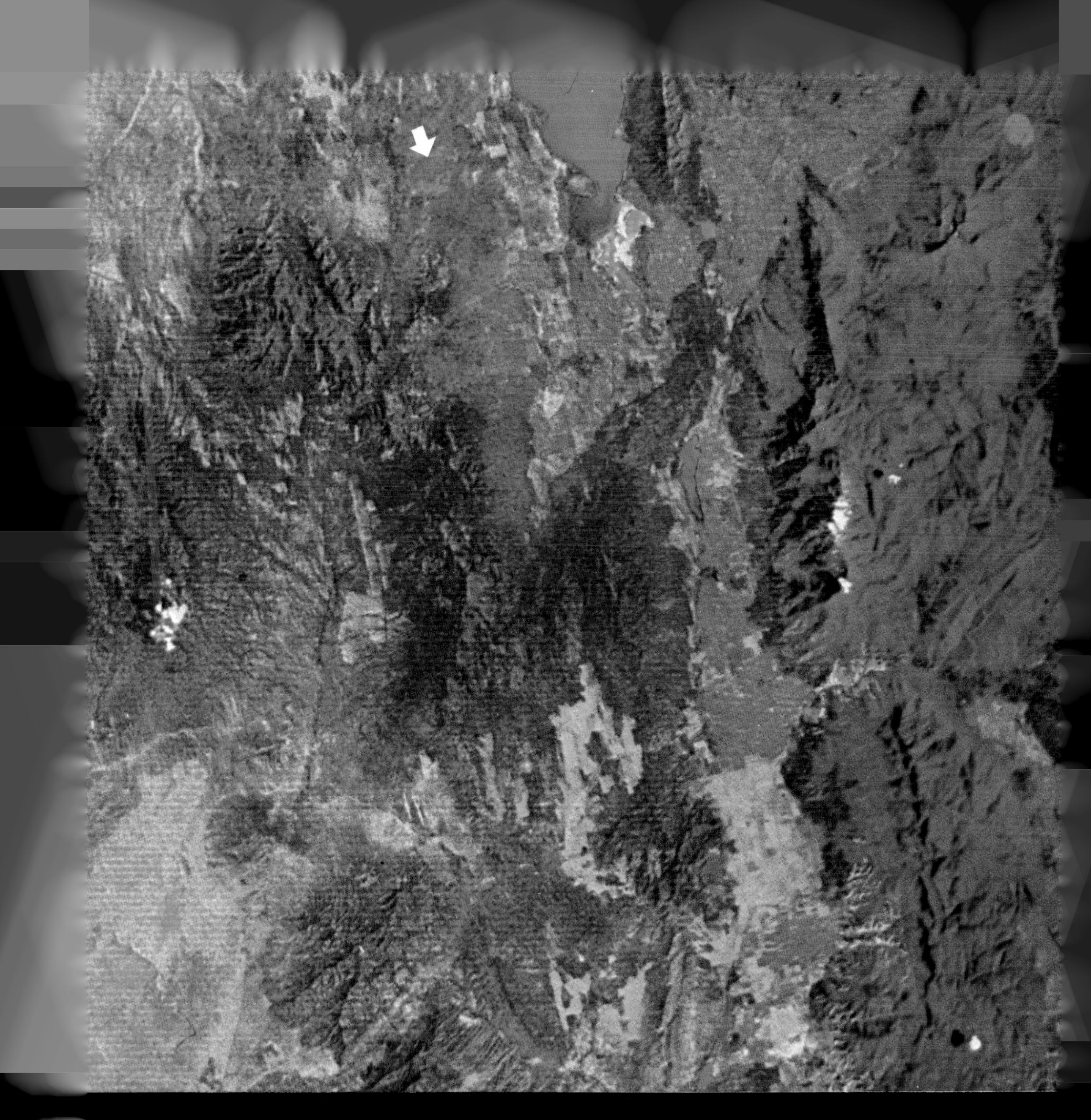

MINERALS

These pictures show two of the target areas used by NASA geologists to establish the ground truth of Landsat images. The sites are used to test for metal ores because in both cases the ground is covered with old mine workings. The tailings and spoil heaps left behind provide a whole palette of spectral signatures.

The intense, glaring colors may seem at odds with those on the previous page, but they are just another way of "stretching" the electronic code. The sense of the colors is not changed – red, for example, still represents the radiation reflected from healthy plants – but the hues have been saturated to show up the slightest variations.

ABOVE: **This is Bingham, Utah, just south of Salt Lake City and famous for the largest hole in the world – the opencast workings of the Bingham Canyon Copper Mine (arrowed), 2,280 feet (1,425m) deep, covering more than 2 square miles. Apart from copper, the area is also rich in gold, silver, zinc and lead.**

The red area at the right of the picture is vegetation on the slopes of the Wasatch Mountains. In the center the trees and shrubs give way to the red and yellow patchwork of fields in the valley.

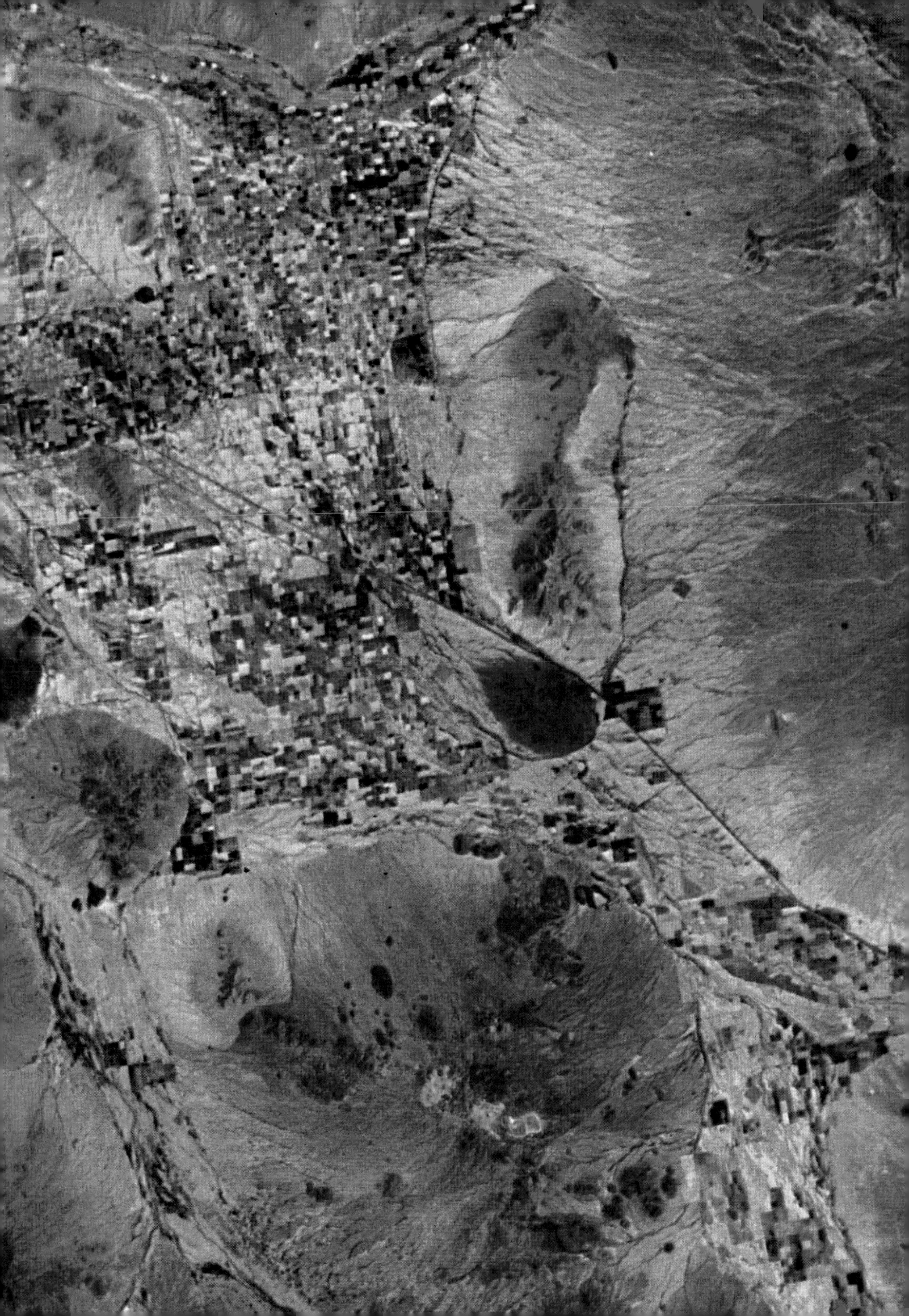

ABOVE: The riches of the Australian outback. On the surface, this part of the Hamersley Range, near the coast of Western Australia, is typical outback country – treeless, sparsely populated, a landscape of eroded hills with occasional patches of spinifex and saltgrass. But just below the surface lies one of Australia's great natural resources, and this Landsat image has been enhanced to show the geological processes that produced it.

The lighter areas are "plutons," massive formations of granite that punched through the earth's crust about 2 million years ago. Although granite is resistant to erosion, it is a comparatively light rock and often comes to the surface in this way. It is also comparatively brittle, as evidenced by the lines across the plutons which are cracks caused by weathering. What interests geologists, though, are the darker areas surrounding them.

The pressures caused by the emergence of the plutons melted the rock around them, allowing basalt to well up from the lower levels. As the mixture cooled, it was metamorphosed into rich mineral deposits. As a result of this geological accident, the granite is now floating in what amounts to a sea of iron ore.

OPPOSITE: The southwest corner of Iceland, with the capital Reykjavik. The vivid reflections off the Langjokull and Myrdaljokull ice caps (top center and lower right) dominate the scene, but among the other features are the linear rock formations (center) forming part of the Mid-Atlantic Ridge, which runs through Iceland, and the purple pattern of sediment discharging from the rivers.

The live volcano-island of Surtsey, which rose from the sea in May 1963 and has been growing ever since, is visible at the bottom of the picture.

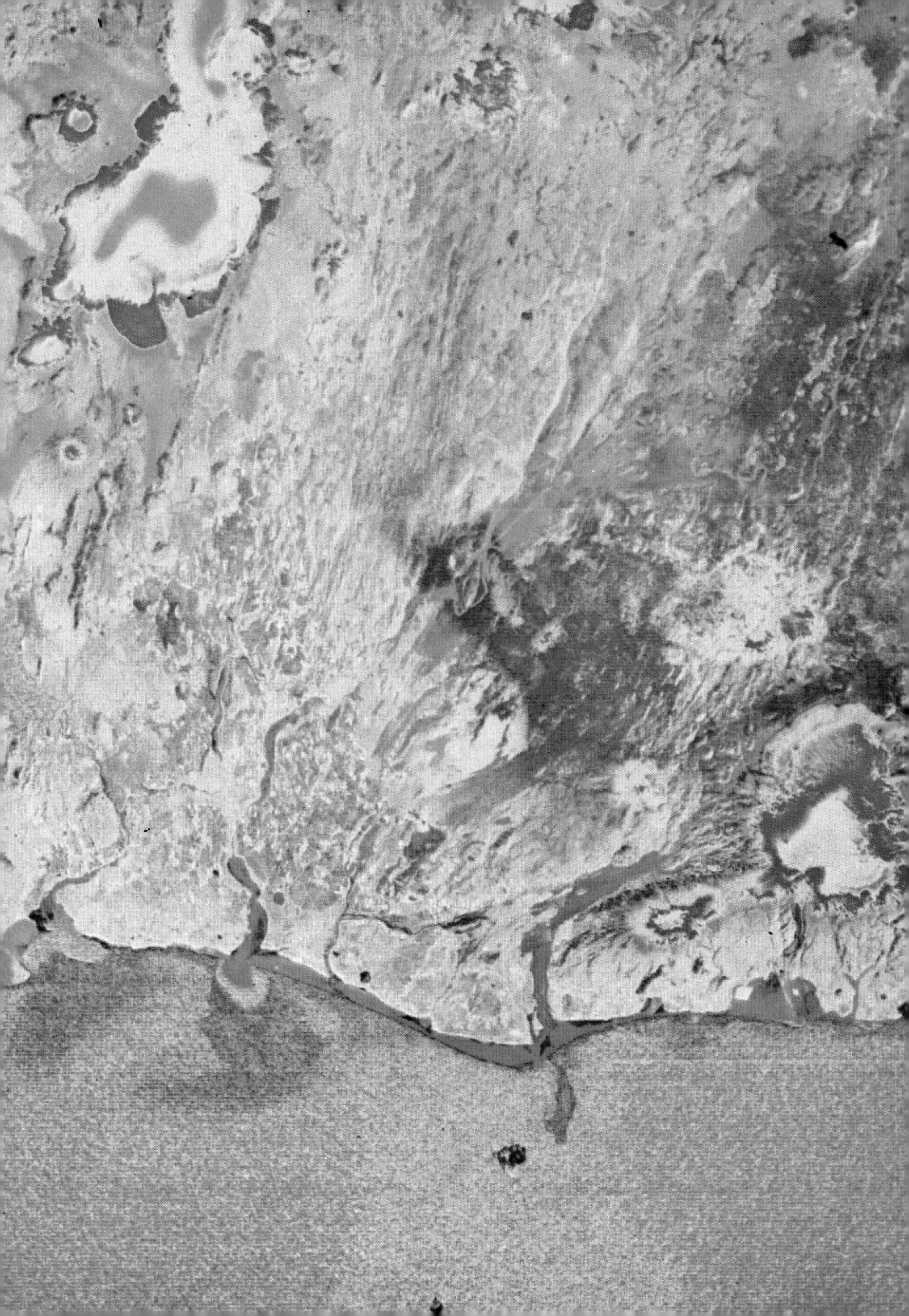

The Appalachian Mountains cascade from north to south like a purple waterfall. Within this dramatic geology – and all but invisible – lie many place names of American history.

The river flowing across the bottom of the picture is the Potomac, while the Susquehanna bends through the top left-hand corner, past the towns of Gettysburg and Harrisburg.

The rivers stand out clearly where they cross the Great Valley of Pennsylvania (left), an extension of the Shenandoah Valley of Virginia to the south. On the far right are the Blue Ridge Mountains, and there is light snow on the Allegheny plateau (top left).

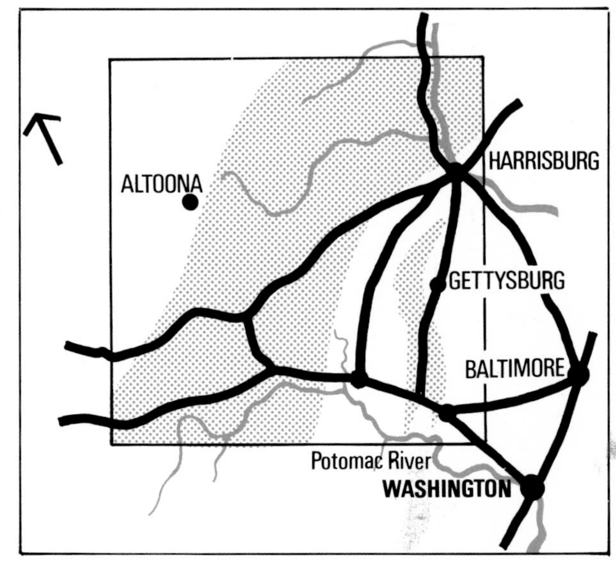

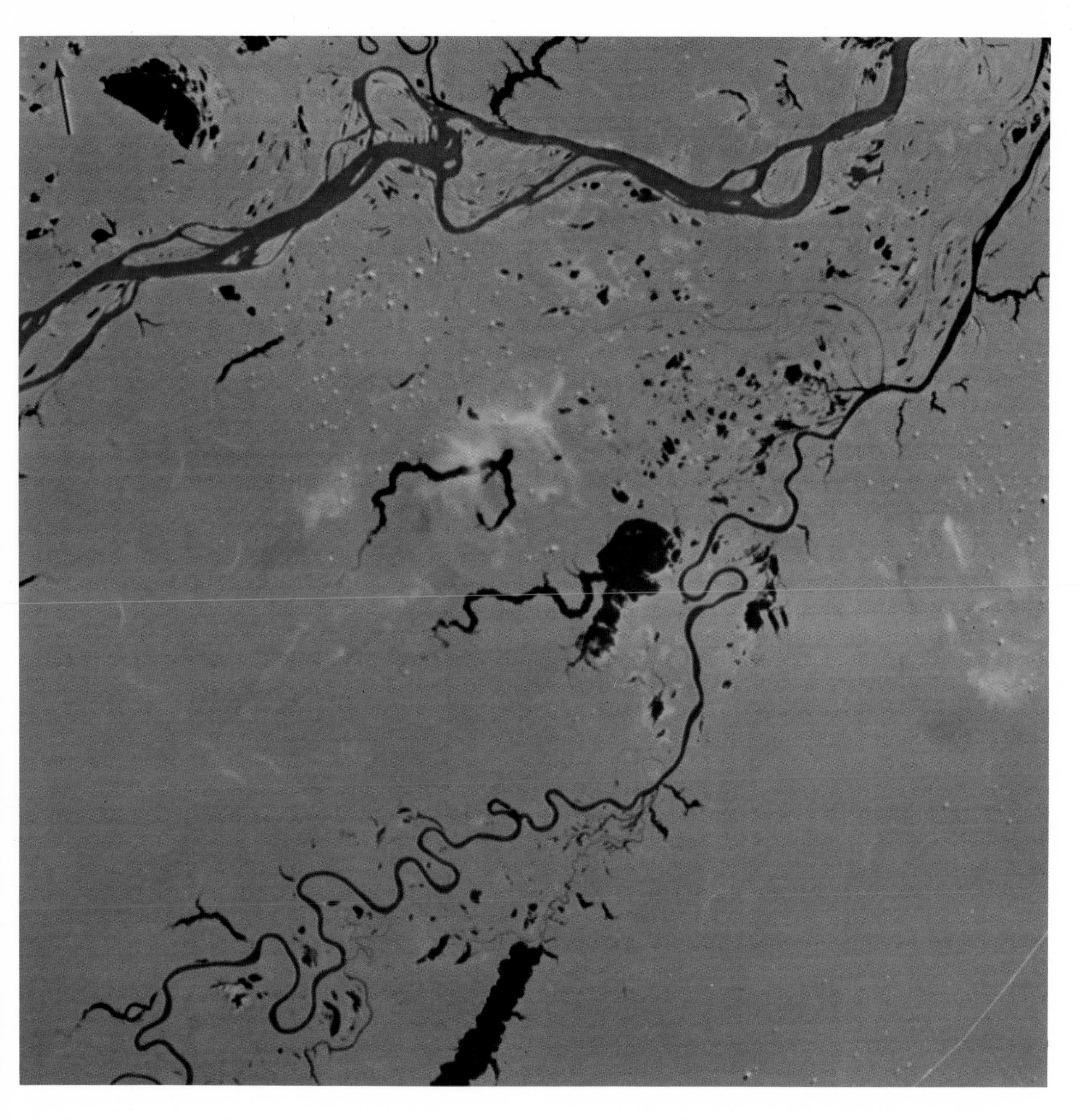

The Amazon basin. This is a lifescape rather than a landscape, because the infrared signature indicates an unbroken carpet of living organisms for hundreds of miles, obliterating every surface feature except for lakes and rivers. The organisms are, of course, the flora and fauna of the tropical rain forest.

Though Lake Aiapua (center) is halfway upstream and approximately 960 miles (1,600km) inland, it is only 100 feet (30m) above sea level. The Amazon is the flattest, as well as the largest, river system on earth. Like the Ganges or any other slow-moving river, one might expect it to be subject to flooding, but the extensive lakes and tributaries, such as the Purus (right), absorb the run-off from the Andes.

The dark color of the tributary shows that the water is relatively fresh. The Amazon itself is so loaded with silt that it registers as blue.

ABOVE: **The Himalayas, from nearly sea level to the peak of Mount Everest, at the top edge of the picture (right).**

The Himalayas are the result of the massive continental collision that occurred when India drifted north from the coast of Africa and slammed into the bottom of Asia. The thrust fault is still rising as India continues to force itself under the northern continent at a rate of 5 inches (125mm) a year.

The eastern half of Nepal covers most of the picture, with Tibet to the north and the Indian frontier running across the plains to the south.

Landsat recorded the view in the "dry" season, so the plains show very little vegetation and the dense bamboo forests in the foothills are laced with empty river beds. During the monsoon rains they become swollen tributaries of the River Ganges. Beyond the foothills are the Lesser Himalayas, rising to 10,000 feet (3,000m). And then the main ranges culminating in Mount Everest itself at 29,028 feet (8,848m).

Most of the Everest expeditions have set out on foot from the Nepalese capital, Katmandu, in its mountain-ringed valley (left). From there it is an arduous three-week march to the base camp at the Thyangboche Monastery just below the mountain, with a final approach up the Imja Khola Glacier.

A shadow lies across the steep North Col, which claimed many lives on earlier expeditions, and a ridge can be seen running south to the companion peak of Lhotse 1, 27,890 feet (8,501m).

OPPOSITE: **The coast of Australia, with sunlight silvering the surface of Port Phillip Bay and Melbourne, visible at the bottom of the picture. Taken from Skylab with a hand-held Hasselblad, the photograph illustrates a natural form of "enhancement" called sunglint, often used in space photography. Although the glare of reflected sunlight tends to obscure ground features, it exaggerates the surface details of water, so that the direction of the swell and even currents show up clearly.**

LEFT: **Clouds build up along the Pacific coast but San Francisco looks across the clear waters of the Bay to Oakland on the eastern shore.**

The road bridge which joins them, via Treasure Island, can just be seen. The Golden Gate Bridge is immediately north of the city and the small speck to the right of it is the prison island of Alcatraz. The notorious San Andreas fault, which has flattened the city once and still holds it under threat of destruction, runs right through the headland, its line marked by the elongated shape of the Crystal Springs lakes. Real estate developments, on landfill, project into the Bay on the left and there are large salt-evaporating basins in the southeast corner.

OPPOSITE: **Perhaps the most historic lake in the world, the Dead Sea. The River Jordan still flows into it from the north, with the traditional site of Jesus' baptism a few miles upstream. Most of the biblical sites are concealed within the rugged folds of the Judean hills, but Jerusalem is visible as a vague blur, halfway between the northern end of the Dead Sea and the edge of the page.**

The picture stops just short of the Israeli coast (left) and the Sea of Galilee (top). The red coloring down the left side indicates intensive agricultural development but, then as now, the shores of the Dead Sea are a parched desert 1308 feet (399m) below sea level. In fact, the sea was slightly shallower and smaller in biblical times, finishing at the bottleneck below the brooding fortress of Massada (on the left shoreline). The southern section is only a yard or so deep. The blue lines are huge evaporation pans, where potash and other chemicals are extracted.

The mountain caves where the famous Dead Sea Scrolls were discovered overlook the northernmost bay on the left side.

THE LAST FRONTIER

The line across the bottom of the picture is one of the few political boundaries visible from space – the edge of the Third world. To the north are the farms and citrus groves of Imperial Valley, California; to the south are the sparsely cultivated cotton-fields of Mexico.

Like the Dead Sea, the valley is an arid desert below sea level, and the glowing checkerboard of fields on the American side of the border was created and is sustained by technology. The entire development, which includes the improbable oasis of Palm Springs with its lawns and golf courses further up the valley and more than half a million acres of farmland, depends on one of the largest irrigation schemes in the western hemisphere.

Even the lake, the Salton Sea, is a recent addition, created at the beginning of the century as an accidental result of attempts to divert the Colorado River. The flood course down which the water spilled can still be seen fanning out from the San Bernadino Mountains on the right. The lake is too salty to be used for irrigation purposes, but fresh water is carried to the valley along the All-American Canal, which runs parallel to the border (bottom right) within sight of Mexico.

The contrast on either side of this "resource frontier" may be less obvious in years to come, as Mexico's new oil wealth enables it to improve its standards of living and agricultural techniques. But Mexico is an exception. For other Third World countries the contrast becomes more desperate every year.

WHOLE EARTH CATALOG

A complete scan by Landsat covers an area roughly 115 miles (185km) square, but it is in these detailed enlargements that the program comes into its own as an earth resources survey.

Each element, or pixel, in the pictures is approximately an acre – the smallest area that Landsat can "see." The average image contains 8 million pixels but, as the value and location of each one is listed, they can easily be counted. Areas, no matter how irregular, can be measured and percentages calculated at the touch of a button. Once the ground truth of each spectral signature is known, the results are remarkably accurate, and Landsat can repeat the survey as often as required, at eighteen-day intervals.

The pictures (LEFT) show 81 square miles (225 sq km) of Williams County, North Dakota. The lighter color represents fields of wheat and small grains (such as barley and oats) which have been picked out by their spectral signature from the surrounding countryside.

When the background information is added again (BELOW), it is possible to see the relationship to geographical features such as lakes and rivers (left), while the intervening pattern of grass and fallow fields now shows up in light and dark greens. Surveys of this sort, carried out on a national scale, provide useful information about the country's agricultural resources and the balance of crops.

Satellites can be used to monitor the local ecology, as in this picture (TOP RIGHT) of the Swift River Reservoir Watershed, west of Richmond, Virginia. In this case the Virginia State Water Quality Board used Landsat data to check whether local construction work was polluting their reservoir. The water in the reservoir is coded blue and the residential areas are red. The gardens and agricultural plots around them are yellow, and the checkered background is a forest of hardwood (light green) and pines (dark green).

Town planners can now use satellite images, like this one (BOTTOM RIGHT) of Tokyo, Japan, to update their maps.

The picture covers about 325 square miles (840 sq km) and the uses of buildings in the city have been divided between commercial (yellow) and residential (purple). Car parks, warehouses and other public buildings are orange and the vegetation in public parks is red. Both the Imperial Palace and Tokyo Airport are clearly visible.

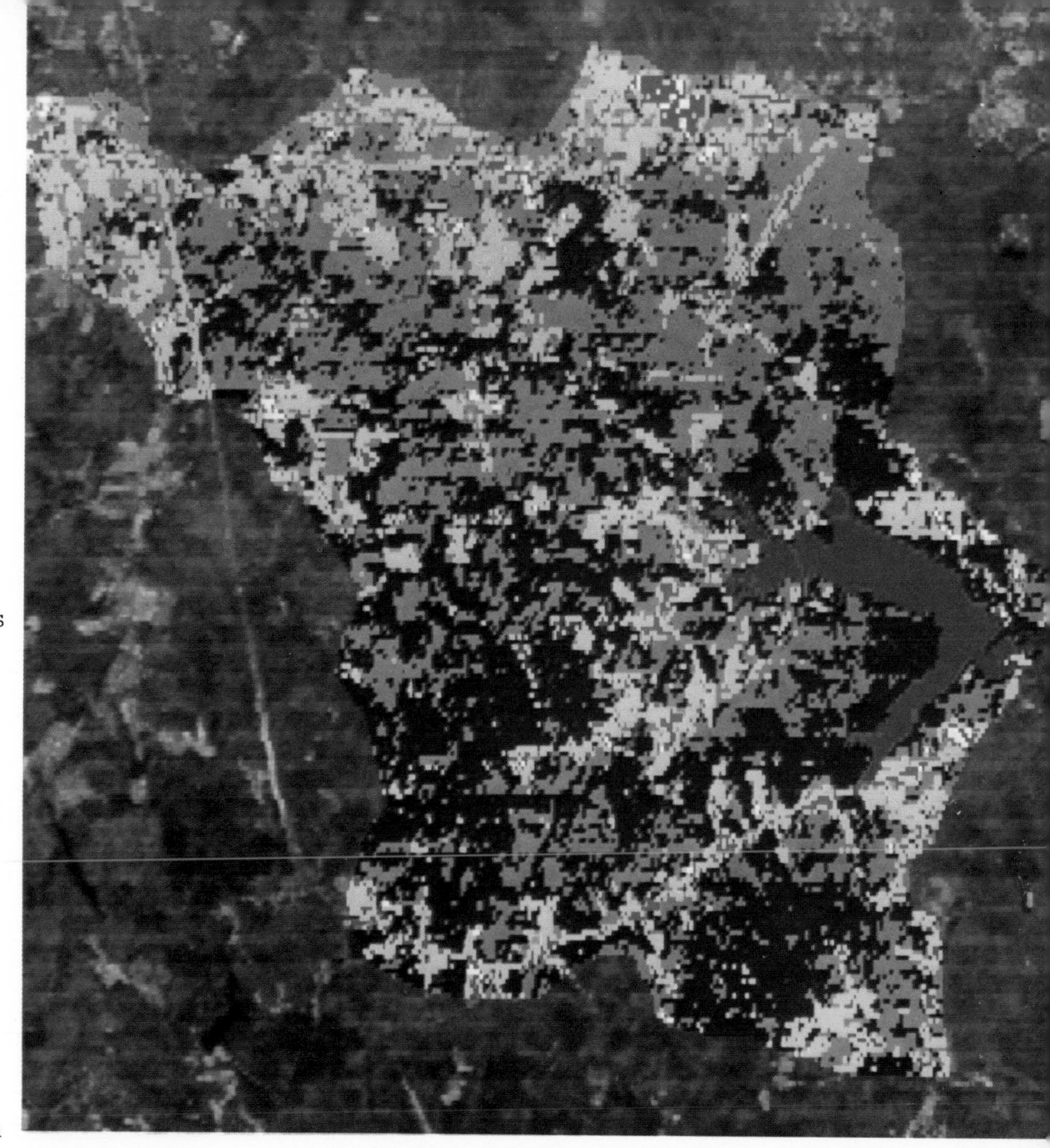

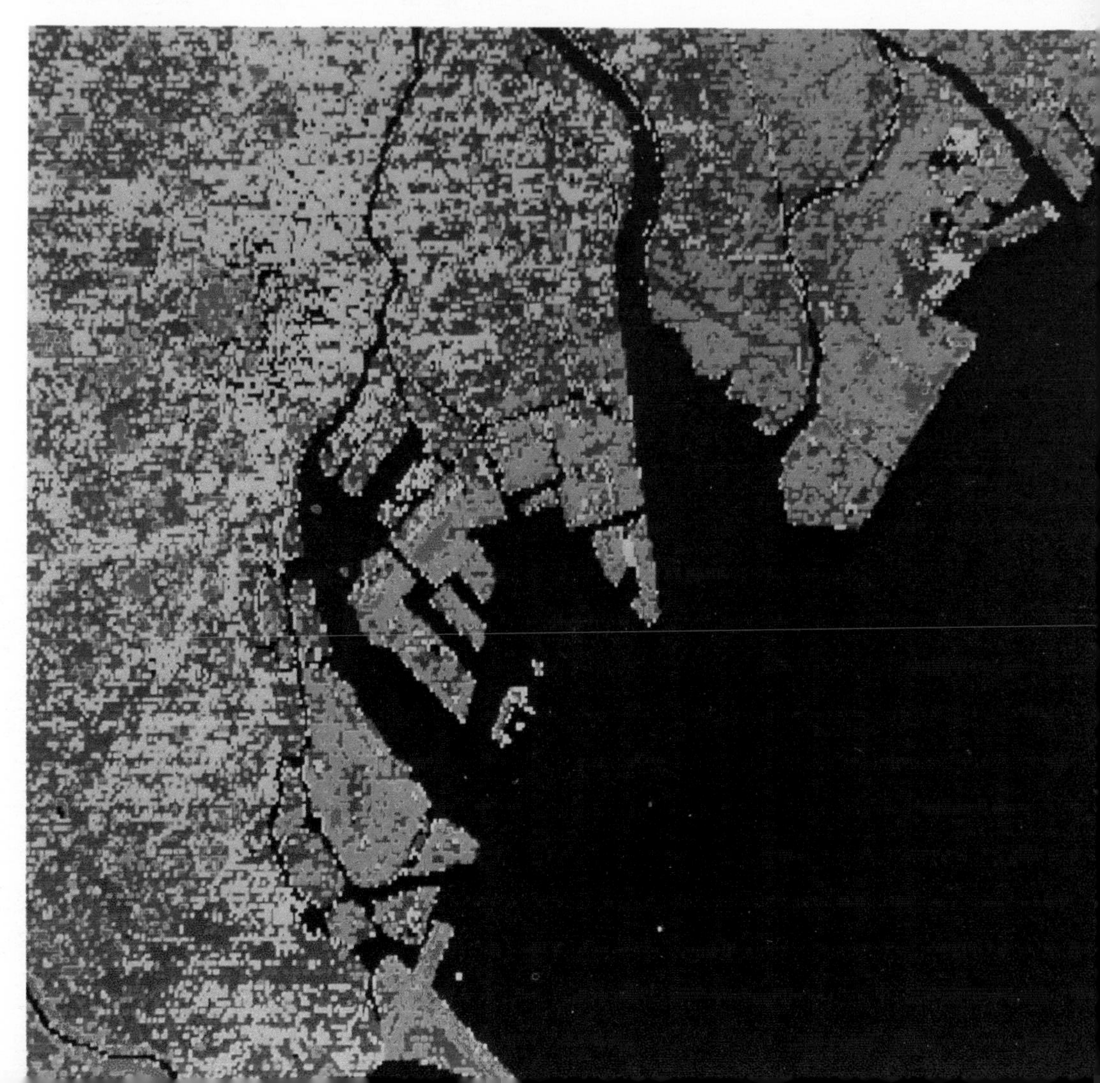

MAPPING THE SEASONS

The city of Denver, Colorado, lies at the foot of America's continental divide, where the grasslands of the "high plains" meet the barrier of the Rockies. The change in geography is emphasized by the seasons and shows the advantage of mapping an area at different times of year.

The seasonal pictures (RIGHT) are early Landsat images, and the enlargement (FAR RIGHT) was taken from the latest satellite in the series. The improvement in quality is not due to the hardware, which is almost identical, but rather to more sophisticated computer programs. The dark rectangular shape northwest of the city is the Rocky Mountain Arsenal (1).

The summer view, in July 1973 (TOP RIGHT), makes use of infrared frequencies to pick out the pattern of vegetation and surface water such as lakes (2). The plains once supported nothing but coarse buffalo grass, but the tributaries of the South Platte River now flow through rich farmland, with fields of wheat (gray) and of barley, oats, corn and vegetables. The mountains are covered with forests of spruce, fir pine and aspen to an elevation of about 11,000 ft (3,300m).

The winter view, recorded in January 1973 (BELOW RIGHT), shows how a light covering of snow can change appearances. The lakes have disappeared, but the mountains are now thrown into three-dimensional relief and the network of roads and urban areas stand out against the white background. The mountains of the Great Divide must have been daunting to the early settlers, but the area is now designated as the Rocky Mountain National Park and interstate highways thread their way up the steep-walled canyons of Clear Creek (3) and over the highest road pass in North America.

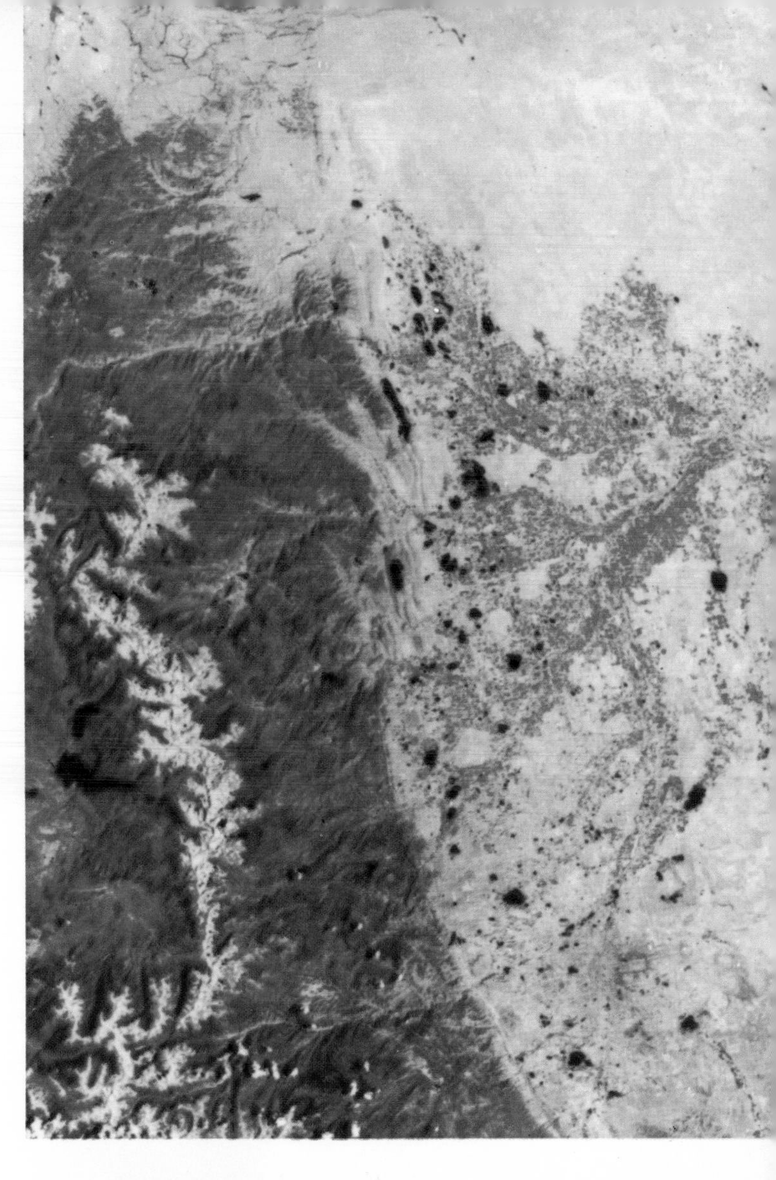

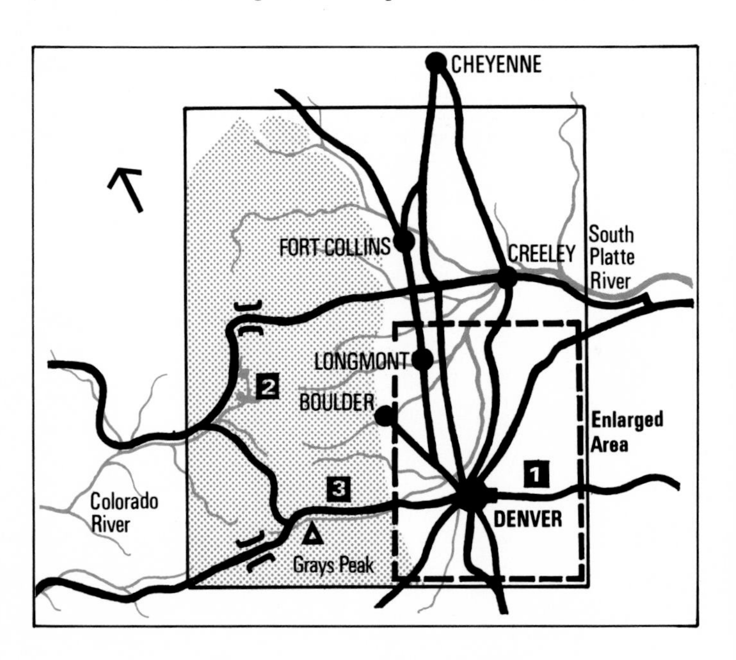

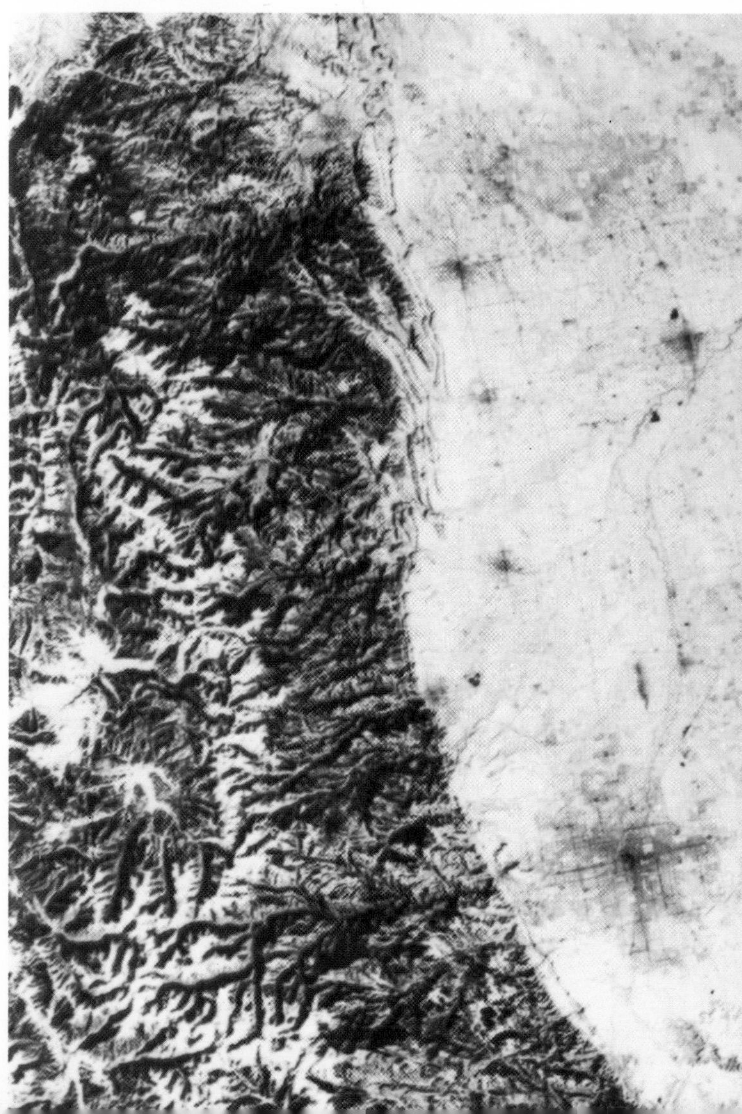

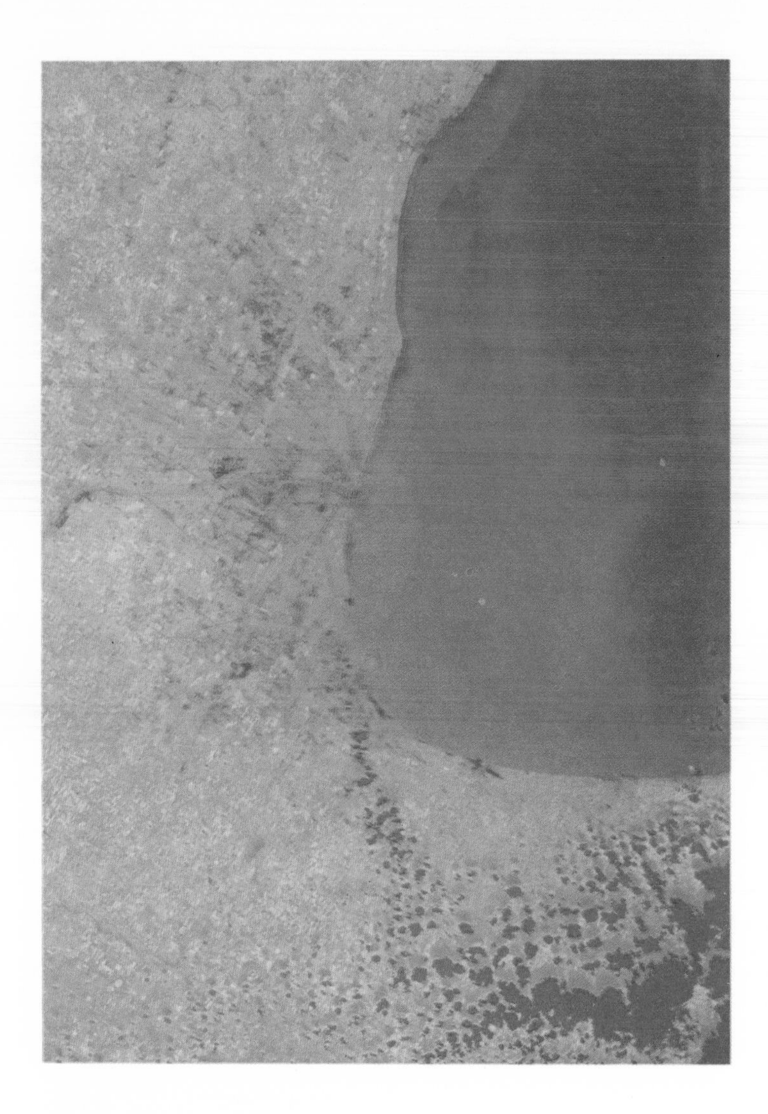

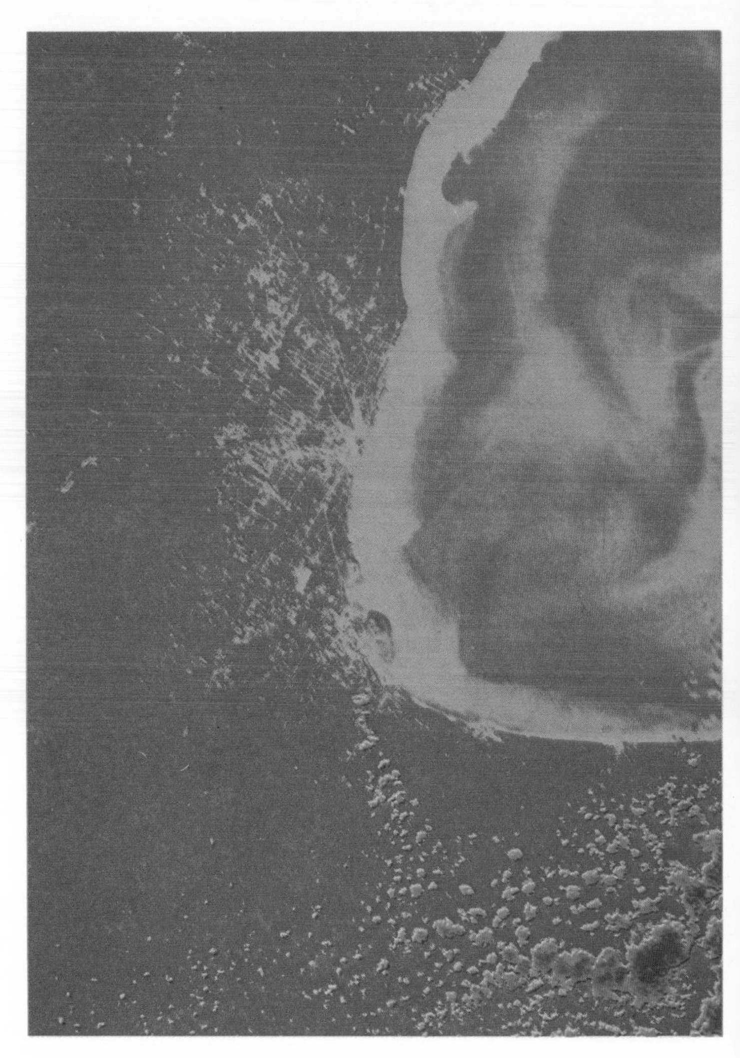

FIRE AND WATER

ABOVE: **Pollution pouring into Lake Michigan from the city of Chicago is recorded by the cameras of Skylab, at a height of 270 miles (432km). The pictures are based on the same infrared photograph, but have been given alternative color codes to show the heavy in-shore pollution** (LEFT) **and the way it disseminates off-shore** (RIGHT). **The survey showed that the pollution was not simply diluted, but remained intact and was caught in swirling patterns miles from the coast.**

Individual particles of waste reflect very little radiation, but water absorbs infrared so completely it provides a black background against which the finest sediment shows up.

At the same time, the picture is a heat map. Warmer, in-shore waters are brightest and the heat from auto engines and shop windows has produced a detailed street map of the city. The colored spots at the bottom of the picture are clouds.

Much of the faint detail has been lost in the upper picture, but it allows the pollution (colored red) to be traced to specific discharge points on the shore. In the lower picture the contrast has been enhanced to bring out the faint details and the stronger readings are bleached out (yellow).

OPPOSITE, ABOVE: **A remote forest fire near Bear Lake, Canada. Fires like this destroy millions of square miles of virgin forest in the Northwest Territories every year, and satellite coverage is invaluable in calculating their direction and spread, and mapping the extent of the destruction.**

The landscape was scoured by prehistoric ice sheets to produce this extraordinary pattern of elongated lakes. The wind is blowing the fire along the path the glaciers once took, with plumes of smoke rising from hot spots along the fire's front. The invisible flames at the back are being blown into the burned-out area.

OPPOSITE, BELOW: **An erupting volcano in the north Pacific punches a hole through the clouds. The volcano is Mount Tiatia in Russia's Kuril Island chain, which runs from Siberia to Japan (left).**

After lying dormant for 161 years, the volcano erupted in 1973 with a noise that was heard 53 miles (85km) away. The plume is seen here rising to a height of 15,000 feet (4,600m) and the eruption has covered one end of the island with smoldering ash.

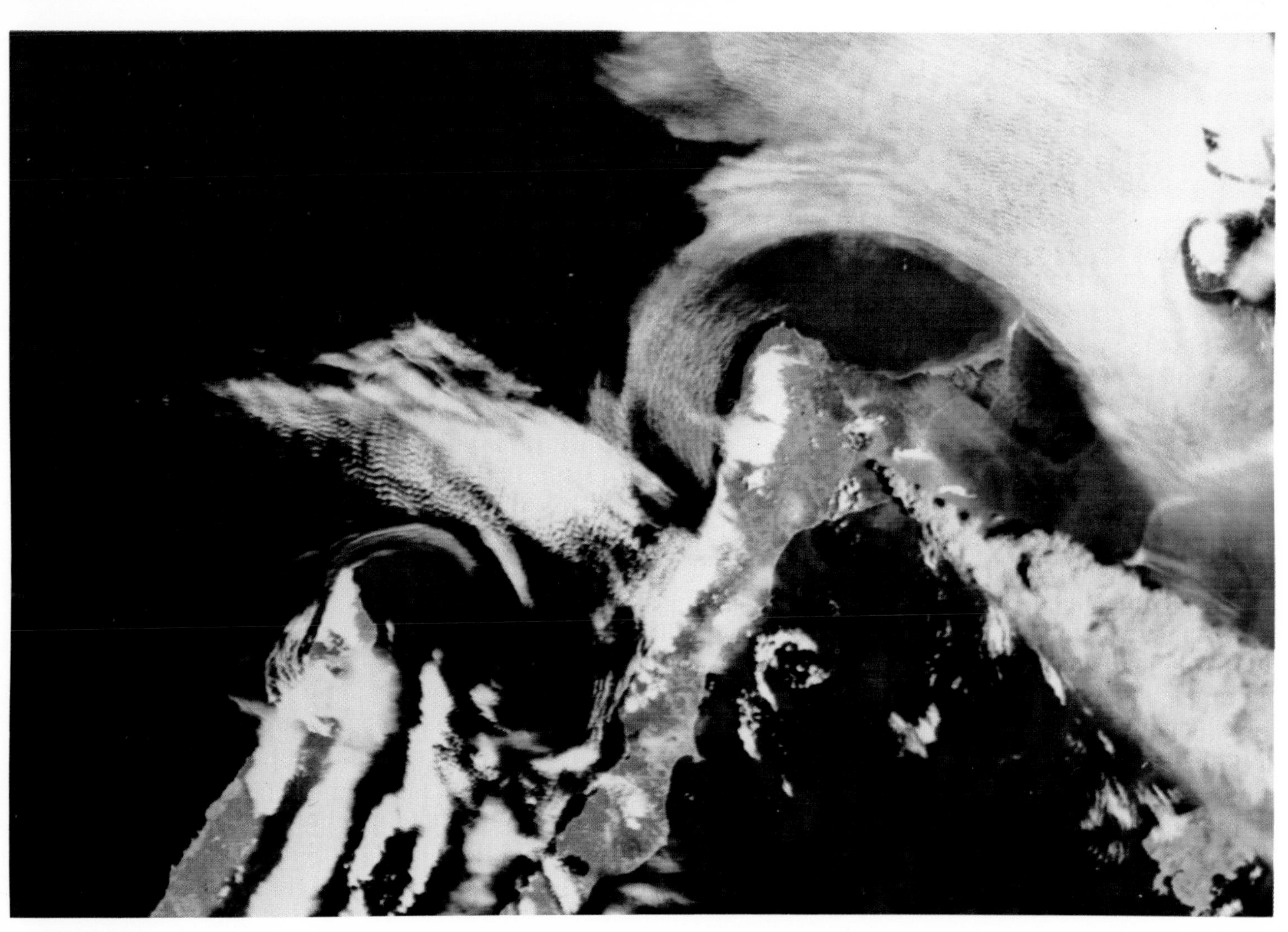

SOUTHEAST ENGLAND

London, with the estuary of the River Thames, recorded by Landsat 1 in March 1973. The picture is slightly distorted, so that the image is stretched horizontally and compressed vertically, but it is possible to make out a wealth of detail across southeastern England.

The color coding includes seventeen shades each of red (vegetation) and green (fallow fields). Coniferous woodlands are dark brown and the lighter browns are deciduous trees.

The nature of the soil varies sharply from the bright greens and yellows of the chalk downs in the south, to the flat lowlands in the north where the peat soil shows up as dark green. The white cliffs of Dover are hidden by cloud, but chalk outcrops along the rest of the south coast stand out in pale yellow.

Apart from the winding course of the Thames itself, it is possible to identify many of the roads and railways radiating from the capital, especially the sharp vertical line of the London to Brighton railway.

Other features of interest include the pattern of runways at London Airport (1), the race courses and paddocks around the horse-breeding center at Newmarket (2) and the earthworks of the New Bedford River, which cuts a white scar across the Fens (3).

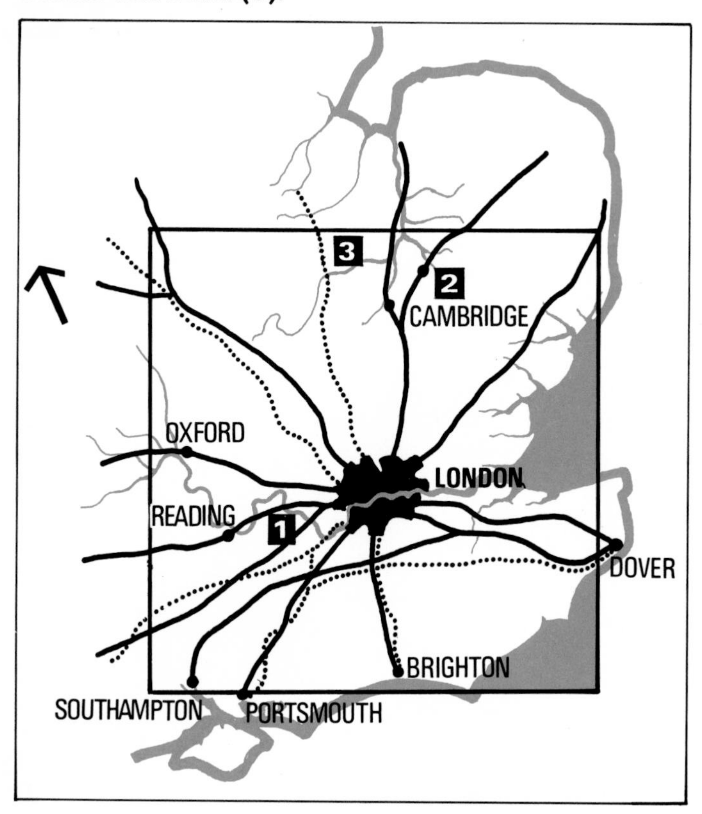

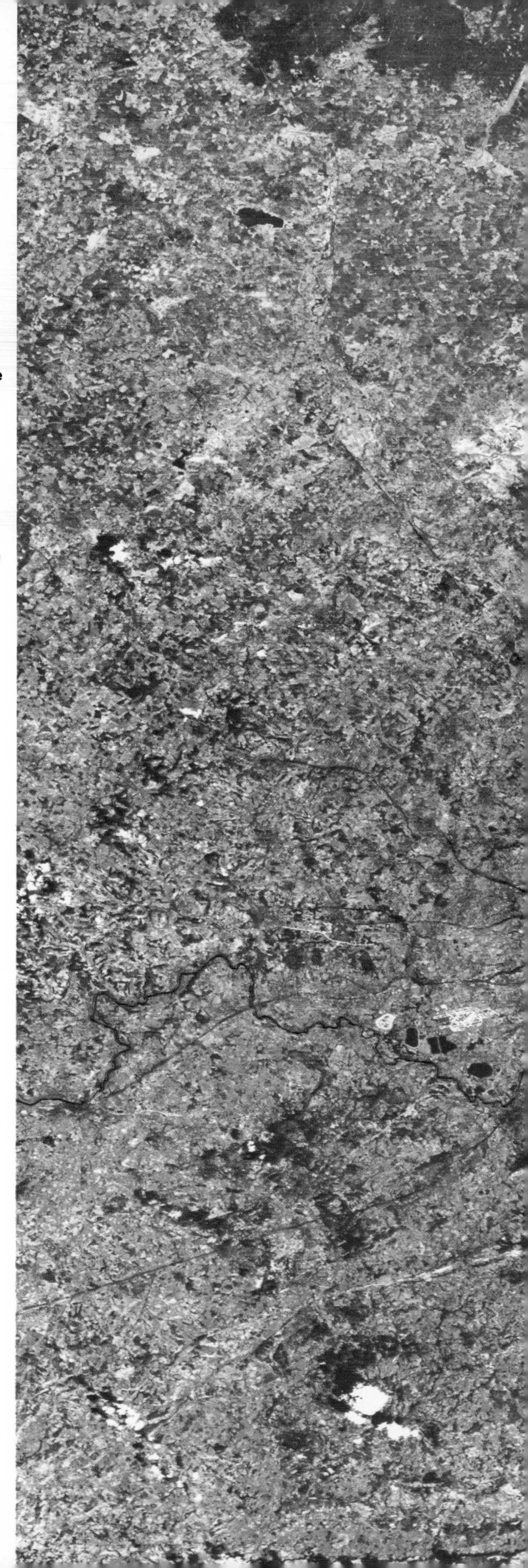

THE CITIES

OPPOSITE: **New York in the spring. A haze hangs over the downtown area, but the beaches of Long Island and the New Jersey coast bask in the May sunshine. The parks, including Central Park and Prospect Park in Brooklyn, stand out as islands of red in the gray built-up area, with their lakes appearing as black dots. The piers and docks of Bayonne, New Jersey, can be seen below the tip of Manhattan Island, where the Hudson and East River flow into the Upper Bay. The sheltered waters of Long Island Sound (top right), and the sand bars which protect its Atlantic coast, have made Long Island a major recreational area for New Yorkers. Jamaica Bay, with its numerous islands, is at the southern end, with Fire Island just along the coast to the right.**

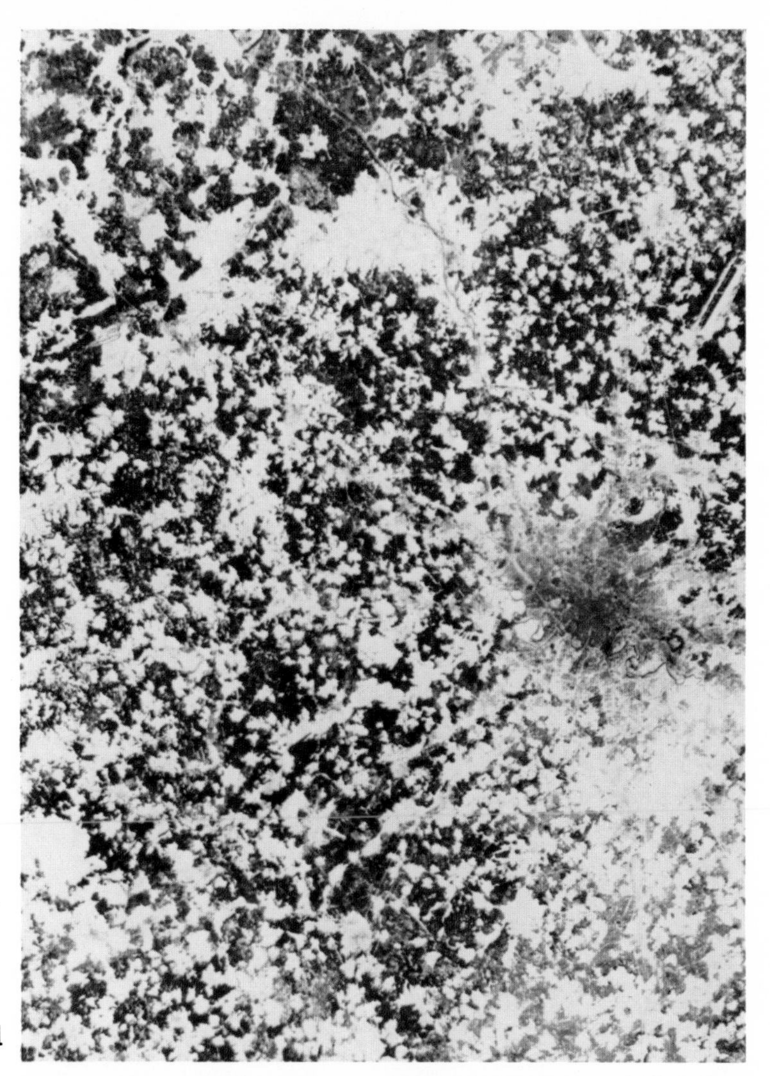

TOP RIGHT: **Moscow in winter. The modern capital of the Soviet Union lies under deep snow cover of the sort that defeated Napoleon's army in 1812. The checkerboard pattern is caused by the mixture of deciduous and evergreen trees, while the city itself is a "heat island" where the snow has melted or been cleared away. Urban environments like this are now known to create their own micro-climate, as much as 5°C warmer than the surrounding countryside.**

The Moskva River flows from east to west through the city and the Moscow Canal is visible above and to the left. This 80 mile (130km) canal eventually joins the headwaters of the Volga River (top left) to form one of the world's most important waterways, linking Moscow to the Caspian Sea.

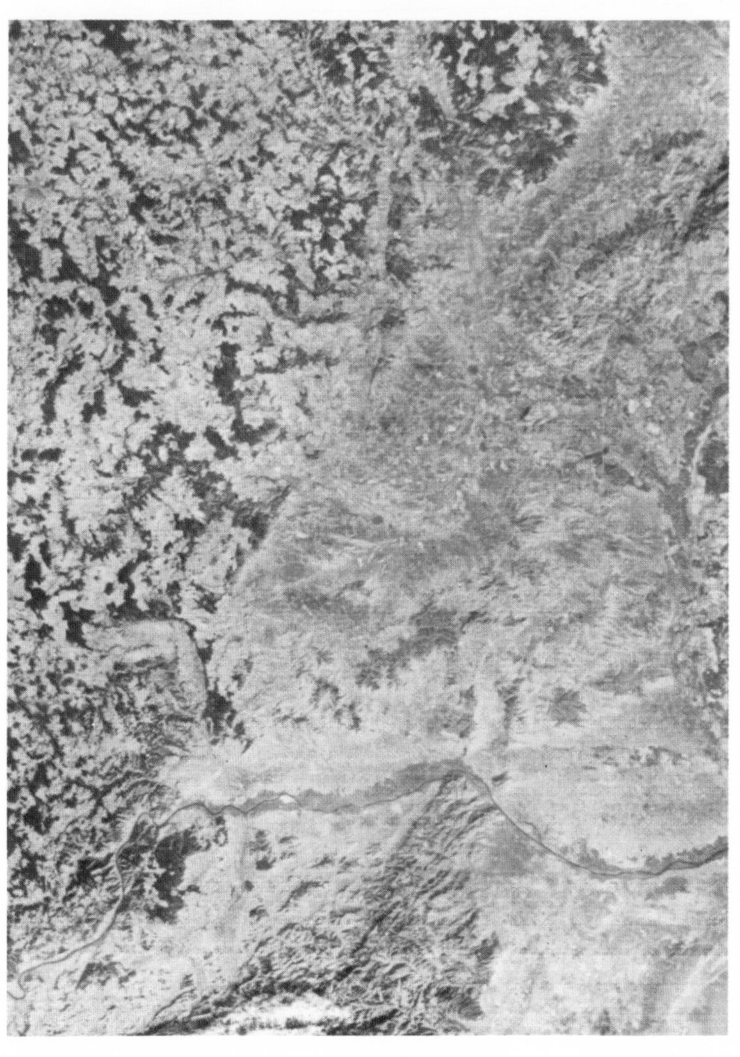

BOTTOM RIGHT: **Vienna in the fall. The city which was the romantic capital of the Austro-Hungarian Empire lies on the River Danube, where it loops around the extreme eastern end of the Alps (blue area, bottom right). In its heyday the forest slopes just outside the city were a fashionable park, made famous in waltz and operetta as the Vienna Woods.**

A century of politics has redrawn the maps of Europe, and the Iron Curtain now runs across the center of the picture. To the south is Austria, to the north is Czechoslovakia, with the town of Brno on a tributary of the Danube (top left).

The river is shared by no less than eight countries on its 1,770-mile (2,800km) journey from Germany to the Black Sea. Downstream lie the capitals of Yugoslavia (Belgrade) and Hungary (Budapest) and for a short stretch it even forms the frontier of Russia.

The mountainous area on the left side is the Bohemian massif and the edge of the Carpathian Mountains appear in the top-right corner. The area is agriculturally poor and most of the vegetation in the picture is forest or woodland.

THE INFRARED WINDOW

This is an infrared image of Mobile, Alabama, taken by the S-109B terrain camera on Skylab. The main purpose was to analyze the silting-up processes in Mobile Bay, but the picture reveals a whole cross-section of human activity.

The area is full of natural resources. Mobile is surrounded by pine forests and the sandy islands along the coast support a wide variety of resort beaches and wildlife refuges. But the forests have made the area a center of the wood-pulp industry and a manufacturing base for paper and rayon; while the sand bars have created such a perfect natural harbor that it is one of the busiest ports on the Gulf Coast, with extensive shipyards and repair facilities.

The Navy has always been here. In fact, a famous naval battle of the Civil War took place within the bay itself, under the batteries of Fort Gaines (on the end of the sand bar, left). The last of the US battleships, the Alabama, is now moored in the bay as a national memorial, and the military presence is reinforced by the Pensacola Navy Air Station and the Brookley Air Force Base, whose runways are visible to the southeast of the city.

A number of transport systems can be seen, including the Interstate 10 to Florida (heading east from the city) and the Intracoastal Canal, joining Bon Secour Bay and Perdido Bay (inside the sand bars, left).

The delicate balance between this concentrated technology and the environment is reflected in the waters of the bay. Any alteration to the flow of silt through the complex estuary of the Mobile River will redistribute the shoals in the bay. Pollution by oil or chemicals, and even waste heat in run-offs, can alter the whole ecology. Each new channel or discharge changes the picture and any disturbance to the waters outside the bay can affect the coast line many miles away.

Through the infrared window we can watch the tracery of currents respond to our uses and abuses, like an analog of the system itself.

THE CONTINENTS

There are few signs, from space, of intelligent life on earth. In broad daylight the results of human activity are no more than faint graffiti on the surface. At night, though, our fires burn like signals.

ABOVE: **Europe by night. This view from a US Air Force satellite was taken in early evening. The streets are busy in London, Birmingham, Liverpool, Manchester and Dublin. Across the North Sea, the lights of Amsterdam, Rotterdam and the Hague run together as a single beacon on the coast. The citizens of both Germanies are at work and play. Berlin shines through a thin cloud cover, and the industrial megalopolis of the Ruhr is a continuous arc of light.**

LEFT: **Europe by day. A single wide-angle view from Meteosat 1 on an almost cloud-free day in June 1978.**

ABOVE RIGHT: **America by day. Composed of 569 separate Landsat images, this photomosaic is the first of its kind. It took several months of dodging cloud cover to produce these remarkably clear scans, and one can identify the location of many of the pictures in this section – from San Francisco (on the extreme left), to Mobile Bay on the Gulf Coast. In the east, Long Island points like a finger at New York, with the Appalachians just to the left. The small bump on the east coast of Florida is Cape Canaveral, where Landsat, Skylab, Apollo and many other spacecraft were launched.**

RIGHT: **America by night. Visible evidence of the most concentrated energy consumption on earth. New York and Chicago glow like nebulae; and the coast of Florida is outlined in a blaze of light.**

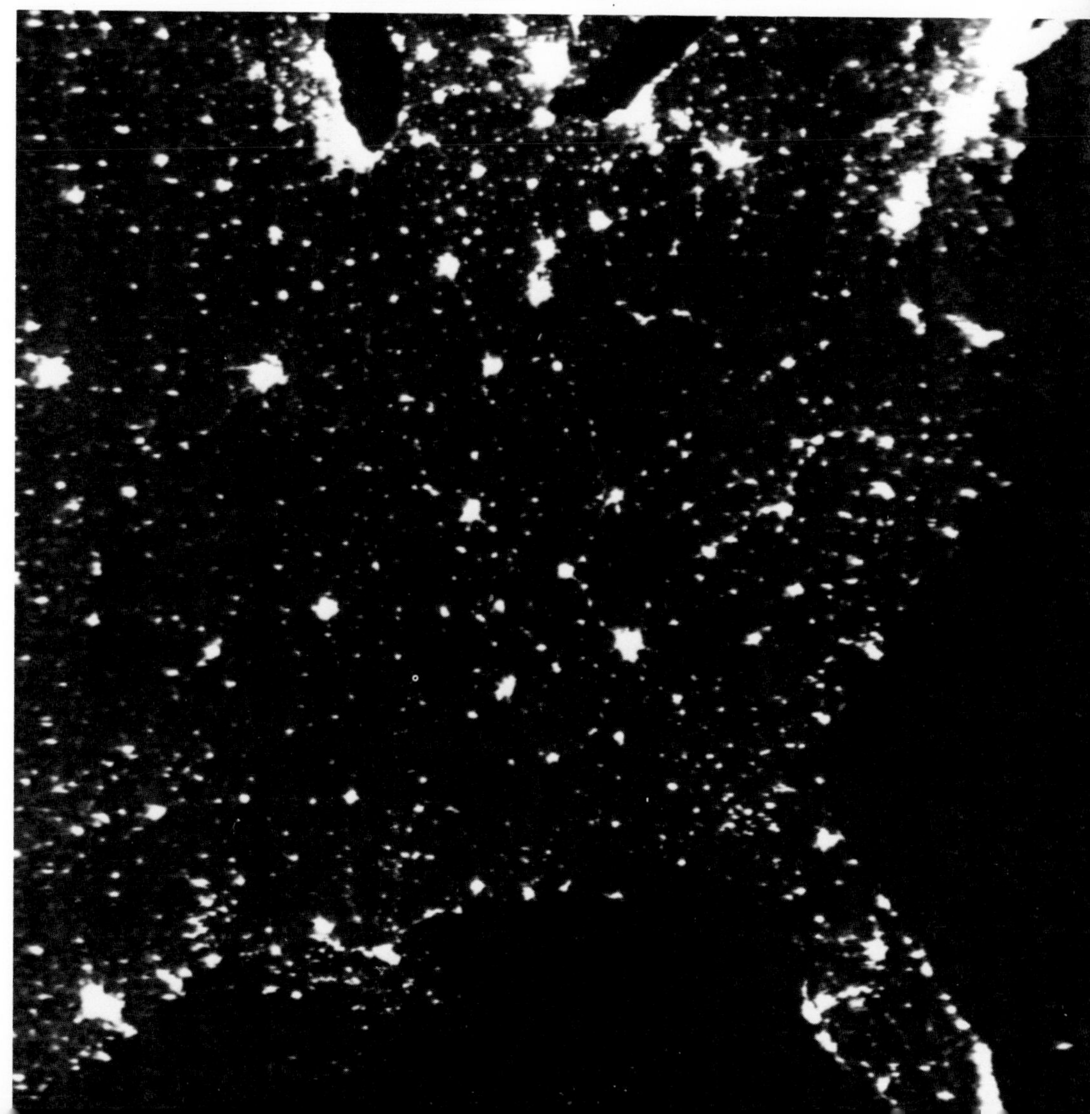

"At the end of the night of time, all things return to my nature; and when the new day of time begins I bring them again into the light."

—*Bhagavad Gita*

Sunset on storm clouds over New Zealand, photographed from Skylab.

INDEX

Page references in italics denote illustrations.